SUMMER AC
PLAYGROUND
WORKBOOK

8 Weeks of Summer Activities

UPDATED FOR 2025 SCHOOL YEAR

15 min a day!

Includes: Math, ELA, Science and Social Studies

GRADE
K-1

Brain Hunter Prep is a division of ArgoPrep dedicated to providing high-quality workbooks for K-8th grade students. We have been awarded multiple awards for our curriculum, books and/or online program. Here are a few of our awards!

Our goal is to make your life easier, so let us know how we can help you by e-mailing us at: info@argoprep.com.

ALL RIGHTS RESERVED
Copyright © 2026 by Brain Hunter Prep

ISBN: 9781951048273
Published by Brain Hunter Prep.

All rights reserved, no part of this book may be reproduced or distributed in any form or by any means without the written permission of Argo Brothers, Inc.
All the materials within are the exclusive property of Argo Brothers, Inc.

Aknowlegments:
Icons made by Freepik, Creaticca Creative Agency, Pixel perfect , Pixel Buddha, Smashicons, Twitter , Good Ware, Smalllikeart, Nikita Golubev, monkik, DinosoftLabs, Icon Pond from www.flaticon.com

- ArgoPrep is a recipient of the prestigious **Mom's Choice Award**.
- ArgoPrep also received the 2019 **Seal of Approval** from Homeschool.com for our award-winning workbooks.
- ArgoPrep was awarded the 2019 **National Parenting Products Award**, **Gold Medal Parent's Choice Award** and a **Brain Child Award**

Summer Activity Playground Series

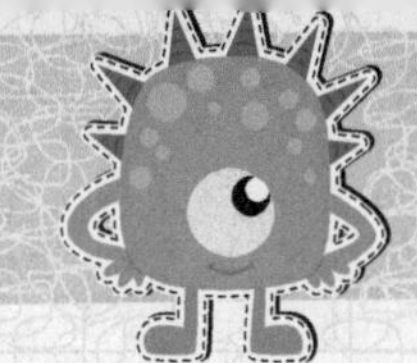

How to use this book?

Welcome to **Summer Activity Playground by Brain Hunter Prep**!

This workbook is specifically designed to prepare students over the summer to get ready for **Grade 1**. Our workbook is divided into eight weeks so students can complete the entire workbook over the summer.

Our workbooks have been carefully designed and crafted by licensed teachers to give students an incredible learning experience. Students will be able to practice mathematics, english activities, science experiments, social studies, and fitness activities. Give your child the education they deserve!

Summer list to read

We strongly encourage students to read several books throughout the summer. Below you will find a recommended summer reading list that we have compiled for students entering into Grade 1.

Author: Jan Thomas
Title: My Toothbrush is Missing

Author: Andy Griffiths
Title: The Big Fat Cow That Goes Kapow

Author: Laurie Keller
Title: We Are Growing

Author: Jon Scieszka
Title: Trucktown

Author: Bonnie Bader
Title: Slow, Slow Sloths

Author: Russell Hoban
Title: Bedtime for Frances

Author: Paulette Bourgeois
Title: Franklin Rides a Bike

Author: Watty Piper
Title: The Little Engine That Could

Author: Audrey Wood
Title: Quick as a Cricket

Author: Jon Scieszka
Title: The True Story of the Three Little Pigs

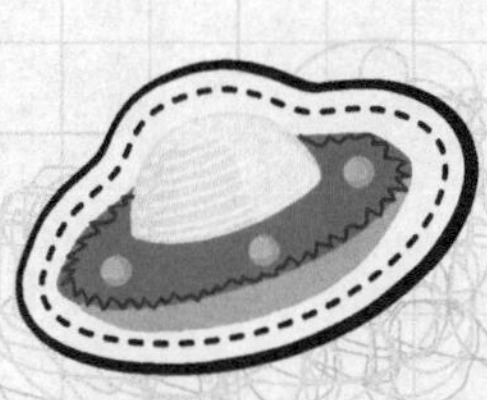

Kindergarten

WEEK 1

Let's get started with:

- counting
- writing numbers
- letters
- rhyming words & more!

1. Count out loud from 1 - 10.

2. Count out loud to fifteen, beginning at the number 6.

3. Count out loud to twenty, beginning at the number 9.

4. Count out loud to 100.

5. Skip count out loud by tens to 100.

6. Skip count out loud by twos to 20.

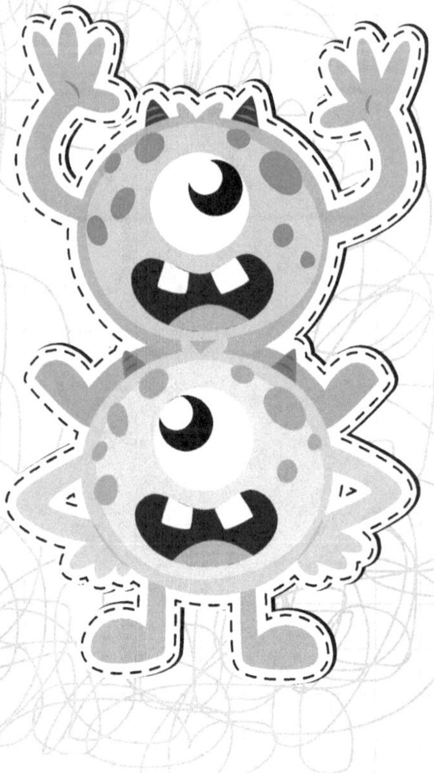

Week 1 Counting & Cardinality

Topic 2 Write Numbers

1. Write the numbers from **1 - 10**.

Practice counting and writing numbers by filling in the missing numbers.

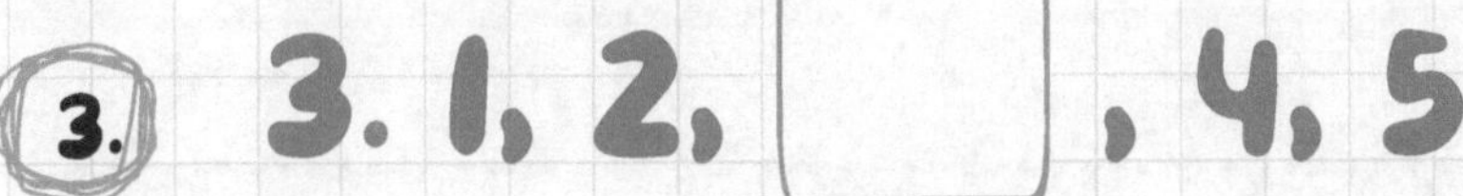

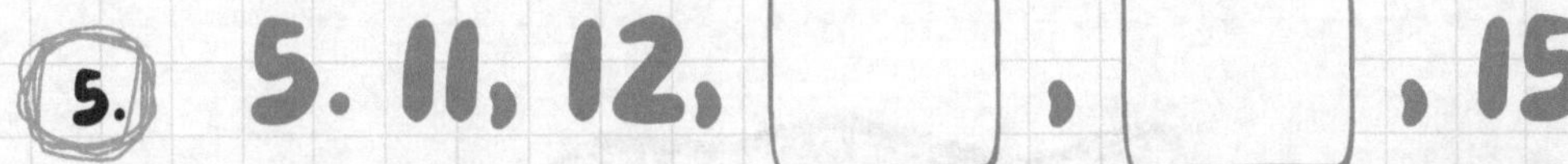

6. 6. 16, 17, ☐ , 19, ☐

Week 1 Handwriting Review

Topic 1 Letters Aa and Bb

Today you are going to review and practice printing the letters Aa and Bb.

Remember **Aa** is for apple and **Bb** is for ball.

Directions: On the lines below, practice printing the letters in your best handwriting.

A ....

a ....

B ....

b ....

Week 1 Phonics Review

Topic 2 Rhyming Words

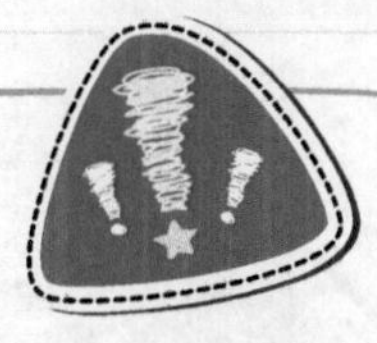

Rhyming words are words that have the same ending sounds. See the list below for examples.

h<u>at</u> and c<u>at</u> m<u>ug</u> and r<u>ug</u> s<u>ip</u> and l<u>ip</u>

Today you are going to practice reading and coming up with your own rhyming words.

Directions: Read each pair of words. Are they rhyming words? Write yes or no on the line.

1. leg and sun → ....
2. rob and sob → ....
3. big and twig → ....
4. bat and mat → ....
5. fun and yum → ....

Directions: Read the word. On the line, write a word that rhymes with the given word.

1. Sit → ....
2. Wet → ....
3. Van → ....
4. Tub → ....
5. Mop → ....

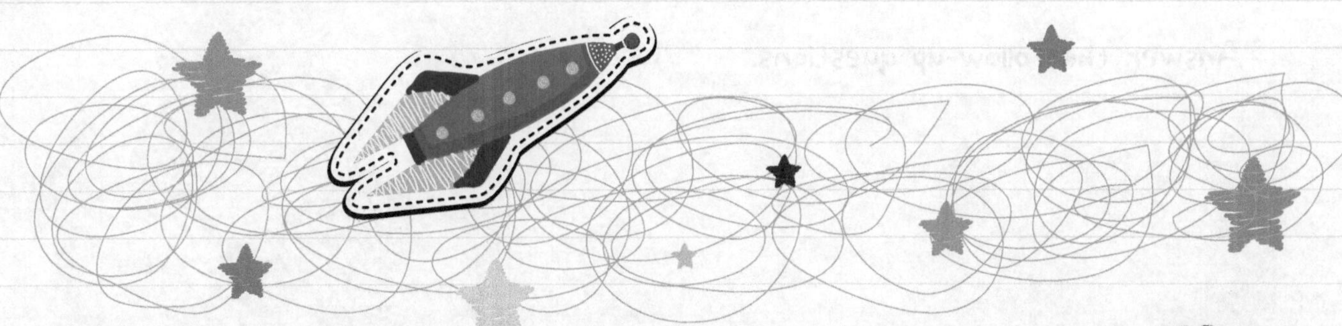

FITNESS PLANET → Let's get some fitness in! Go to page 111 to try some fitness activities.

You can use **observable properties** to sort objects. This means that you can explore the way something looks and feels. Size, color, weight and shape are all observable properties.

Today you will use observable properties to sort objects.

Materials Needed:

* A variety of objects such as dry beans or pasta, paper, toys, coin, ball, cotton ball, etc.

Procedure:

1. Collect 4-5 different objects you want to observe.
2. Observe them using your five senses, according to the properties listed below.
3. Answer the follow-up questions.

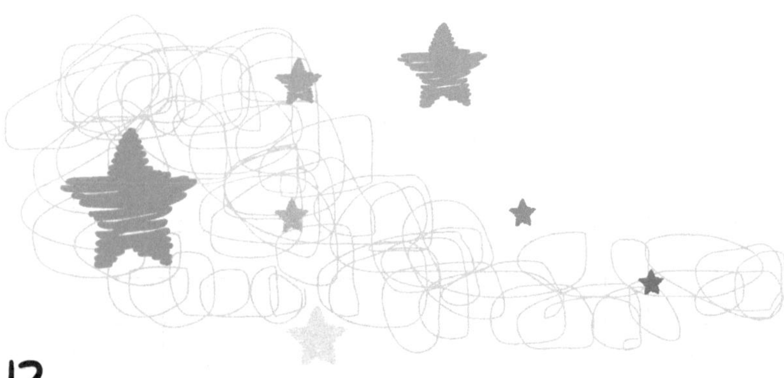

Object	Color	Size	Shape	Weight

Follow-Up Questions

1. Which objects could be sorted as large?

2. Which objects could be sorted as small?

3. Which objects could be sorted as heavy?

4. Which objects could be sorted as light?

5. Which objects are most alike?

Week 1 Counting & Cardinality

Topic 3 Topic: Ordering Numbers

Place the following numbers in the correct order.

1. 

2. 

3. 8, 11, 9, 10, 12 ☐, ☐, ☐, ☐, ☐

Fill in the missing numbers on the number line.

4. 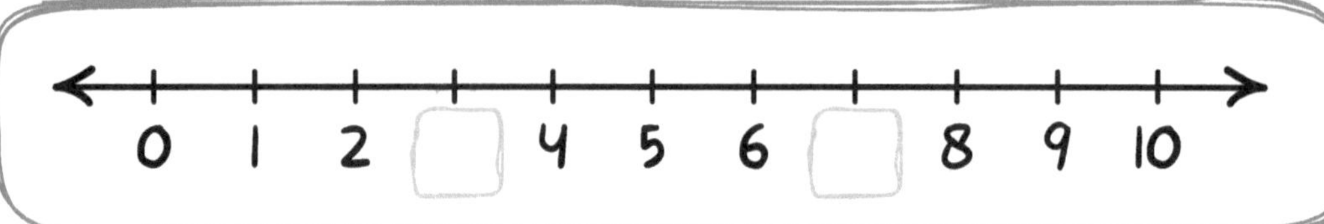

5. 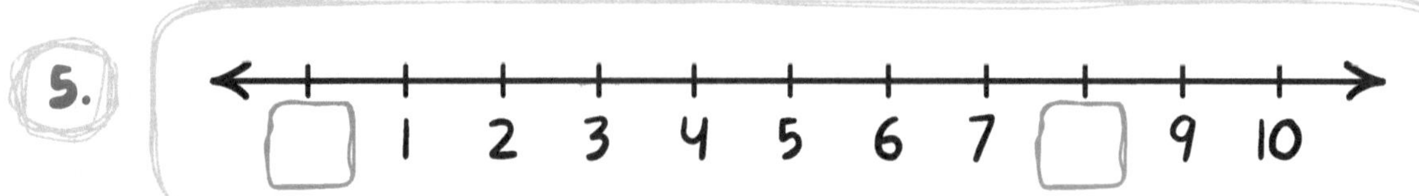

6. 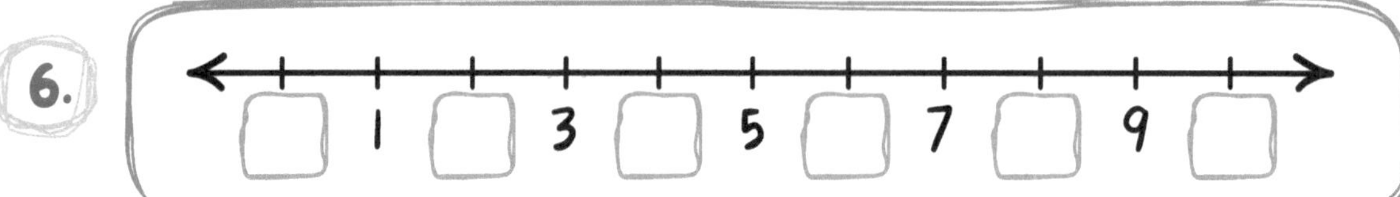

Topic 4 One-to-One

1. Count the objects.

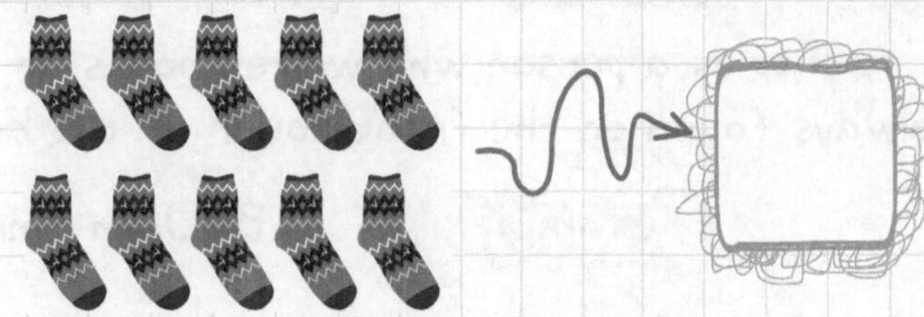

2. Count the objects.

3. Count the objects.

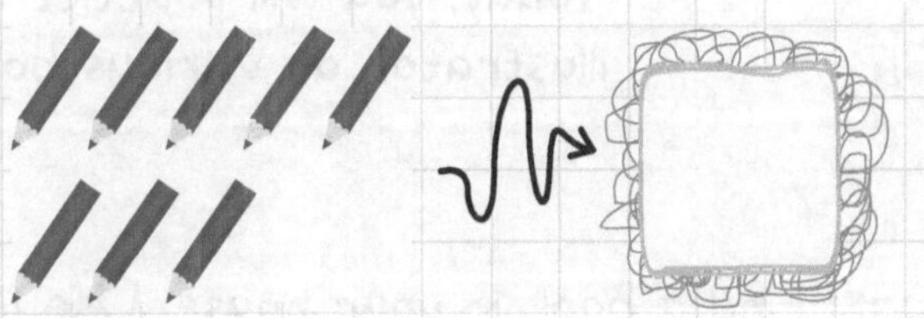

4. How many ducks are there?

5. How many stars are there?

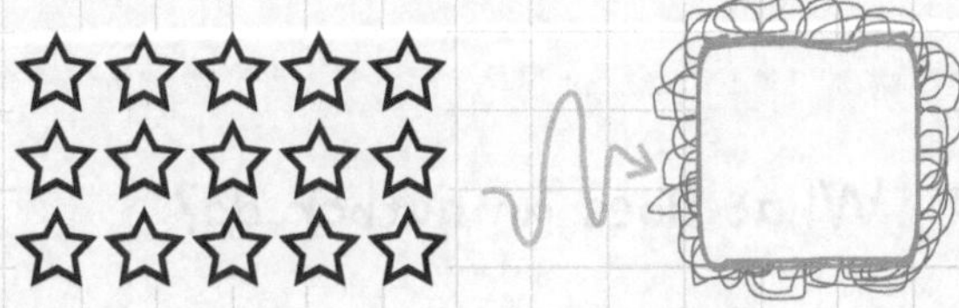

6. How many smiley faces are there?

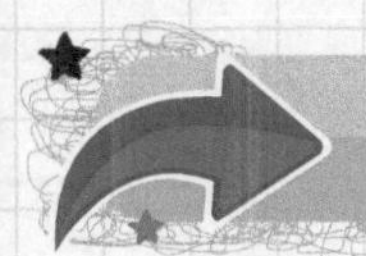

FITNESS PLANET

Let's get some fitness in! Go to page 111 to try some fitness activities.

An **author** is a person who writes books or poems. An author's name is always found on the front cover of the book.

Example: By John Smith

An **illustrator** is a person who creates the pictures in a book. An illustrator's name is usually found on the front cover of the book as well.

Example: Illustrated by Anna Johnson

Today, you will practice identifying the author and illustrator of various books.

Directions: Find a book in your house. Use it to answer the questions below.

1. Book Title: ....

2. Who is the author of the book? ....

3. How do you know this person is the author? ....

4. What does an author do? ....

5. Who is the illustrator of the book? ....

6. How do you know this person is the illustrator? ....

7. What does an illustrator do? ....

Week 1 | Language Review

Topic 4 | Nouns

A **noun** is a person, place or thing.

Examples:

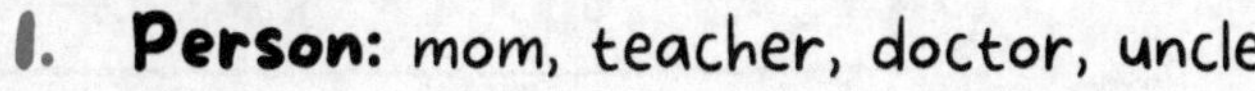

1. **Person:** mom, teacher, doctor, uncle

2. **Place:** home, New York, Wichita, Germany, restaurant
3. **Thing:** ball, dog, bus, bed, pencil

Directions: Answer the questions below.

1. Is *school* a noun?

   A. Yes   **B.** No

2. Is *run* a noun?

   A. Yes   **B.** No

3. Write 3 nouns for each category listed below.

   A. Person
   B. Place
   C. Thing

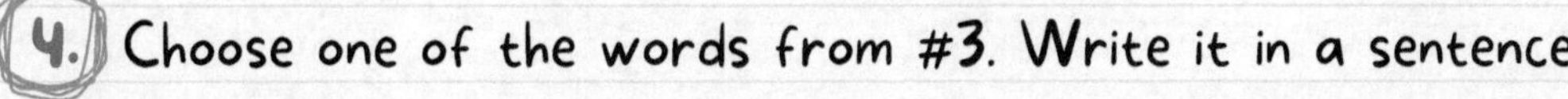

4. Choose one of the words from #3. Write it in a sentence.

....

....

....

....

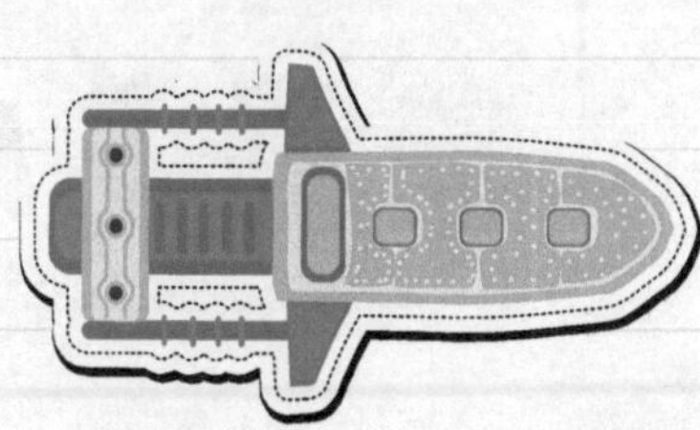

 Circle 10 objects.

 Circle 8 objects.

 Circle 6 paw prints.

 Put an X over 11 fish.

 Put an X over 16 animals.

 Put an X over 19 pieces of clothing.

Topic 1 Children and Families in History

There are many things that are different about children and families today than in the past. Today, you are going to explore these differences.

Directions: Have a family member help you use the internet to find information about what it was like to live in the past. Use this information to help you answer the questions below.

1. Draw a picture of what houses looked like in the past and what they look like now.

Past	Now

2. Draw a picture of what a child might have worn in the past and what they wear now.

Past	Now

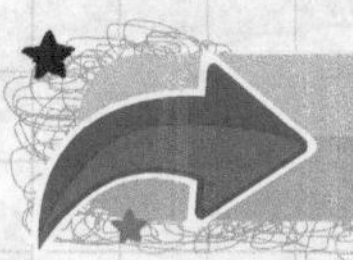

FITNESS PLANET

Let's get some fitness in! Go to page 111 to try some fitness activities.

Kindergarten

WEEK 2

Get ready to learn about:

- ten frames
- comparing numbers
- syllables
- addition and more!

1. How many does the ten frame show?

2. How many does the ten frame show?

3. How many does the ten frame show?

4. Color the ten frame to show the number of objects.

5. Color the ten frame to show the number of objects.

6. Color the ten frame to show the number of objects.

Week 2 Counting & Cardinality

Topic 2 Compare Number of Objects

1. Are there more apples or grapes?

2. Are there more sharks or elephants?

3. Which has less, shoes or socks?

Write the symbol or greater than (>), less than (<), or equal to (=).

Topic 1 Letters Cc and Dd

Today you are going to review and practice printing the letters Cc and Dd.

Remember **Cc** is for cat and **Dd** is for dog.

Directions: On the lines below, practice printing the letters in your best handwriting.

C ....

c ....

D ....

d ....

FITNESS PLANET

Let's get some fitness in! Go to page 111 to try some fitness activities.

All words have chunks of sounds that include a vowel and one or more consonants. These are called **syllables**. Words may have 1 syllable or many. Syllables are sometimes easier to identify by clapping each individual sound heard in a word. Let's take a look at a few examples.

Van	van	This word has 1 syllable.
Playground	play / ground	This word has 2 syllables.
Elephant	el / e / phant	This word has 3 syllables.
Alligator	al / li / ga / tor	This word has 4 syllables.

Now it's your turn to practice.

Directions: Read each word. Count or clap out the syllables. Write the number of syllables on the line.

1. Zebra → ....
2. Vacation → ....
3. Sing → ....
4. December → ....
5. Cafeteria → ....

Directions: Write a word that has the correct number of syllables.

1. 1 syllable ....
2. 2 syllables ....
3. 3 syllables ....

Topic 1 Using Properties to Determine Uses

Last week, you collected and sorted objects by their observable properties, using your five senses. Today, you will brainstorm different ways these objects can be used.

Directions: Choose one of the objects you observed last week to help you answer the questions below.

Object:

1. How do people usually use this object?

....

....

....

....

....

....

2. What is another way the object could be used?

....

....

....

....

....

....

3. Ask a family member at home to do the same thing you did in #2. Record their answer below.

4. How were your answers alike?

5. How were your answers different?

Topic 3 Compare Numbers

1. Which is larger, 17 or 7?

2. Which is smaller, 4 or 8?

3. Which is larger, 6 or 9?

4. Which is smaller, 0 or 3?

5. True or false: $8 > 7$?

6. True or false: $15 < 5$?

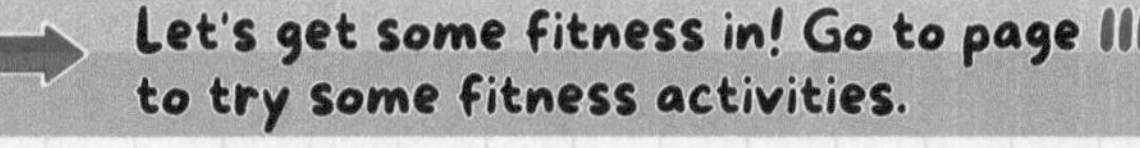
Let's get some fitness in! Go to page 111 to try some fitness activities.

1. 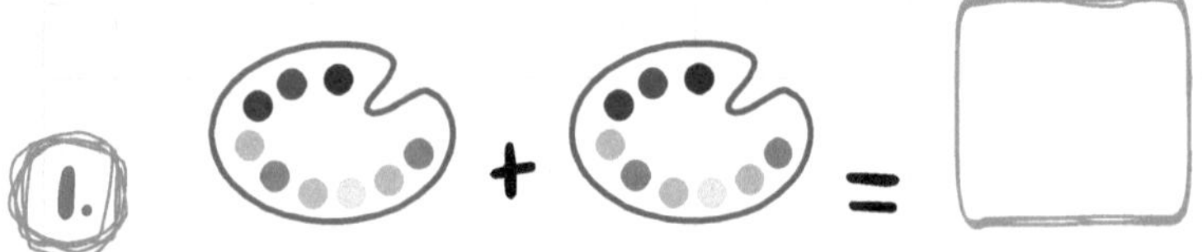

2. 

3. + =

4. + =

5. + =

6. + =

Week 2 Reading Review

Topic 3 Literature

Characters are the people or animals in a story. Characters are who the story is about.

The **setting** is where the story takes place. Books may have more than one setting.

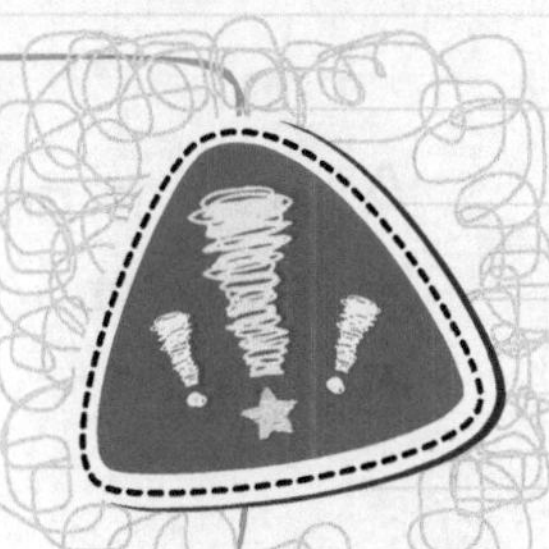

Today, you are going to explore the characters and setting of a book.

Directions: Choose a book from your house. Read it on your own or with a family member. Use it to answer the following questions.

1. Title of book: ....

2. Author: ....

3. Illustrator: ....

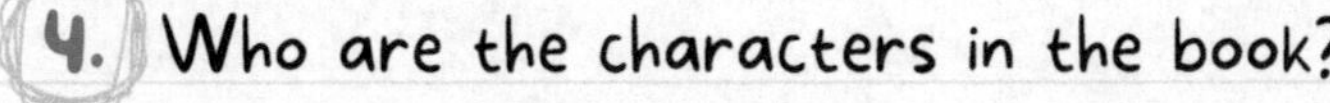

4. Who are the characters in the book? ....

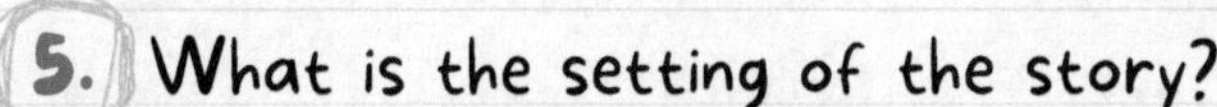

5. What is the setting of the story? ....

6. Did you like this book? ....

Topic 4 Verbs

A **verb** is an action word. It is a word that you can act out.

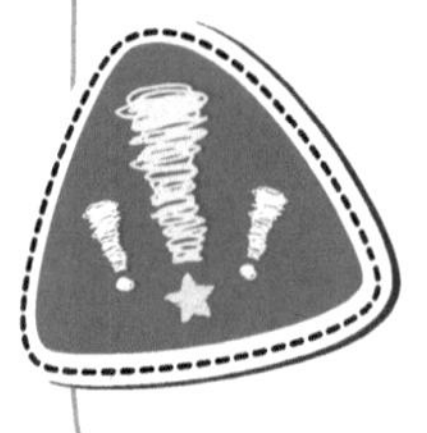

Examples:

1. kick
2. swim
3. play
4. build
5. swing

Directions: Answer the questions below.

1. Is *treehouse* a verb?

   A. Yes **B.** No

2. Is *painting* a verb?

   A. Yes **B.** No

3. Write 5 verbs below.

   A. ....
   B. ....
   C. ....
   D. ....

4. Choose one word from #3. Write a sentence using that word.

....

....

....

....

Topic 5 Operations: Addition (With Representation)

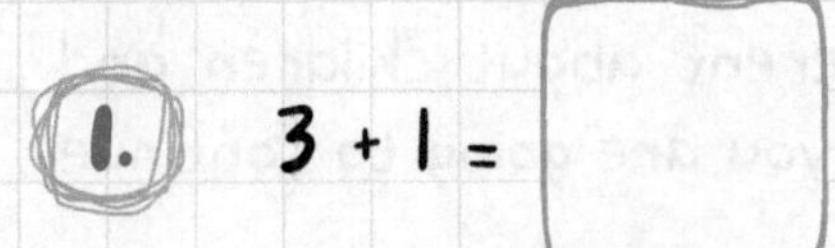

1. 3 + 1 = ☐ (3 pears + 1 pear)

2. 2 + 3 = ☐ (2 snails + 3 snails)

3. 1 + 5 = ☐ (1 bee + 5 bees)

4. 4 + 4 = ☐

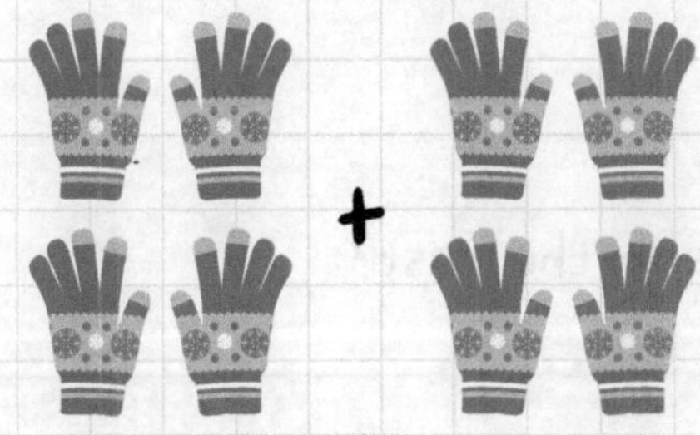

5. 2 + 0 = ...

6. 6 + 5 = ...

FITNESS PLANET

Let's get some fitness in! Go to page 111 to try some fitness activities.

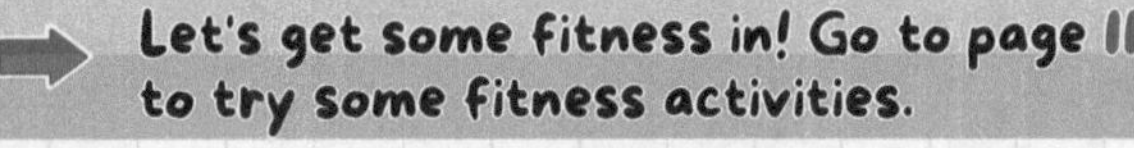

Topic 1 Children and Families in History

There are many things that are different about children and families today than in the past. Today, you are going to continue to explore these differences.

Directions: Have a family member help you use the internet to find information about what it was like to live in the past. Use this information to help you answer the questions below.

1. Draw a picture of how a child might have gotten to school in the past and how they get to school now.

Past	Now

2. What do you think it would have been like to live in the past?

....

....

3. Do you think it was harder or easier to live in the past? Why?

....

....

4. What is one thing you would like about living in the past?

....

....

5. What is one thing you would NOT like about living in the past?

....

....

Kindergarten
WEEK 3

Let's get to work on:

- letters Ee and Ff
- one syllable words
- pushes and pulls
- subtraction and more!

Week 3 Operations & Algebraic Thinking

Topic 1 Operations: Addition within 10

3 + 1 = . . .

5 + 1 = . . .

3. 3 + 4 = . . .

$$\begin{array}{r} 3 \\ +\ 3 \\ \hline \ldots \end{array}$$

$$\begin{array}{r} 6 \\ +\ 4 \\ \hline \ldots \end{array}$$

6.

$$\begin{array}{r} 3 \\ +\ 7 \\ \hline \ldots \end{array}$$

1.

$$\begin{array}{r} 7 \\ +\ 4 \\ \hline \ldots \end{array}$$

2.

$$\begin{array}{r} 9 \\ +\ 5 \\ \hline \ldots \end{array}$$

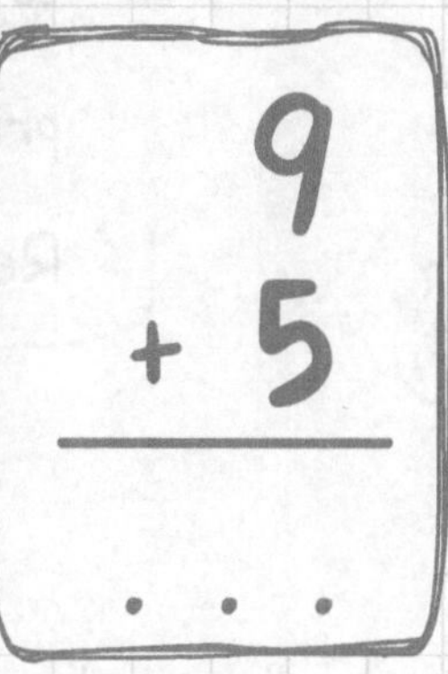

3.

$$\begin{array}{r} 15 \\ +\ 3 \\ \hline \ldots \end{array}$$

4. 12 + 1 = . . .

5. 9 + 9 = . . .

6. 7 + 5 = . . .

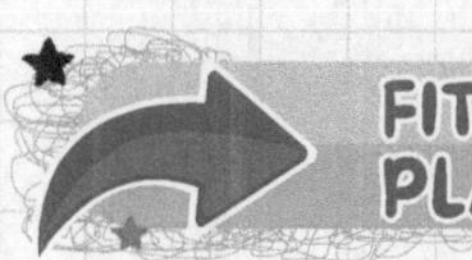

FITNESS PLANET

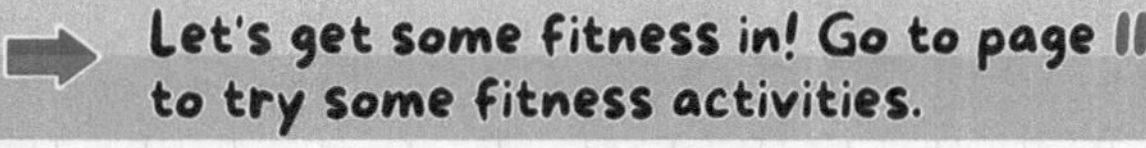

Let's get some fitness in! Go to page 111 to try some fitness activities.

Topic 1 Letters Ee and Ff

Today you are going to review and practice printing the letters Ee and Ff.

Remember **Ee** is for egg and **Ff** is for frog.

Directions: On the lines below, practice printing the letters in your best handwriting.

E ....

e ....

F ....

f ....

Topic 2 Blending One Syllable Words

One syllable words can be sounded out by blending the onset (beginning sound) and rime (middle and ending sounds) together.

Today, you are going to practice blending one syllable words.

Directions: Read each onset and rime below. Blend them together to make the word.

1. r - ug
2. s - ad
3. f - ig
4. b - ed
5. l - og
6. s - eed
7. m - ade
8. r - oad

Directions: Think of 5 one syllable words. Write them below, separated into the onset and rime like in the the words you just read.

1. ....
2. ....
3. ....
4. ....
5. ....

A **push** and a **pull** are both forces that make an object move or stop moving. Pushing an object moves it away from us. Pulling an object moves it closer to us.

With your hands, practice pushing an invisible object away from your body and pulling it back close to your body.

Today, you are going to brainstorm objects that can be pushed or pulled.

Directions: Brainstorm objects that can be pushed or pulled. It may be helpful to think of objects at home or at school. Fill out the chart below. Then answer the questions that follow.

Push	Pull

Follow-Up Questions

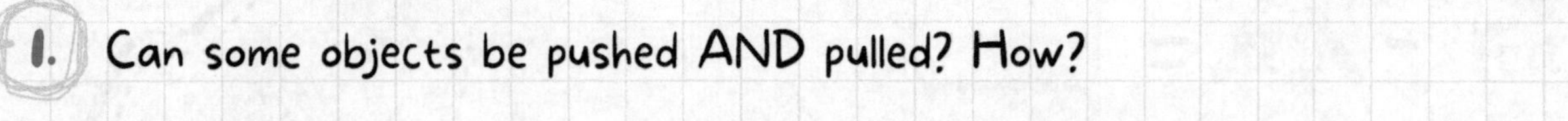

1. Can some objects be pushed AND pulled? How?

2. Give an example of an object that can be pushed AND pulled.

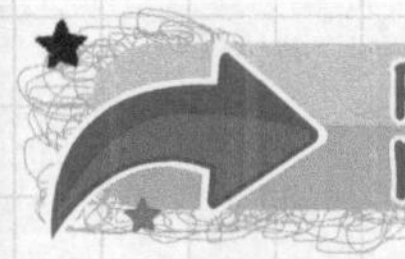

FITNESS PLANET → Let's get some fitness in! Go to page 111 to try some fitness activities.

1. 

2. 

3. 

4. - =

5. - =

6. 

Subtract the crossed-out objects and write the number to solve the problem.

1. 5 - 2 = ☐

2. 5 - 4 = ☐

3. 5 - 1 = ☐

4. 10 - 1 = ☐

5. 10 - 3 = ☐

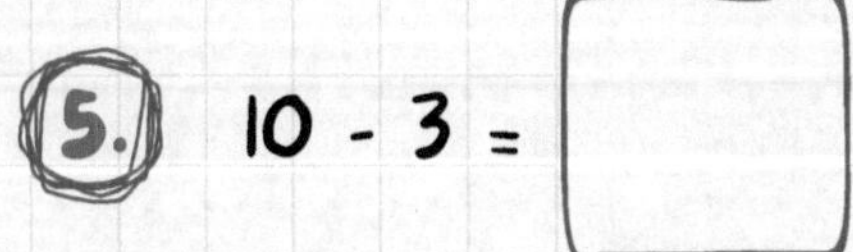

6. 10 - 5 = ☐

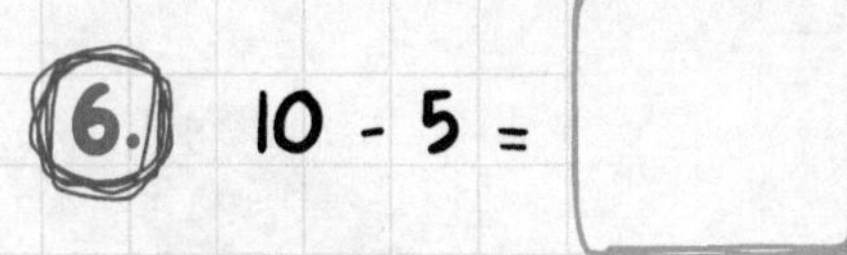

Read the passage below. Then, answer the questions that follow.

Pat is a cat. He likes to lay in the sun all day. Pat also likes to eat food and play with string.

Hank is a dog. He likes to run and fetch balls outside. Hank does not like Pat at all.

Pat and Hank live in the same house. They belong to Jack. Jack loves Pat and Hank, but he wishes they would get along better.

One day, Jack has an idea. He goes to the store and buys a bunny. Jack names the bunny Fluffy. Jack takes Fluffy home to meet Pat and Hank. Pat and Hank do not like Fluffy.

Now, Jack spends a lot of time with Fluffy, and Pat and Hank have no one to play with them. Finally, they are so sad that they start to play with each other. Now Pat and Hank are friends!

1. Who are the characters in the story? ....

2. What is the setting of the story? ....

3. Why do you think Pat and Hank do not like each other? ....

4. What was Jack's idea? ....

5. Why do Pat and Hank decide to become friends? ....

Topic 4 Plural Nouns

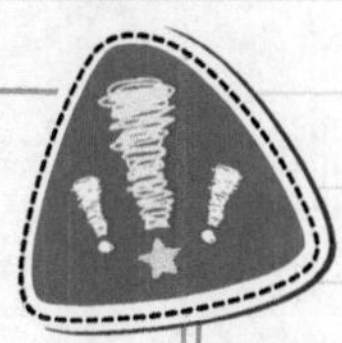

A **noun** is a person, place or thing. When there is more than one person, place or thing, the noun becomes **plural**. Plural means more than one.

To make a noun plural, we add an -s or -es to the end of the word.

Examples:

Teacher → teachers

Tree → trees

Dish → dishes

Box → boxes

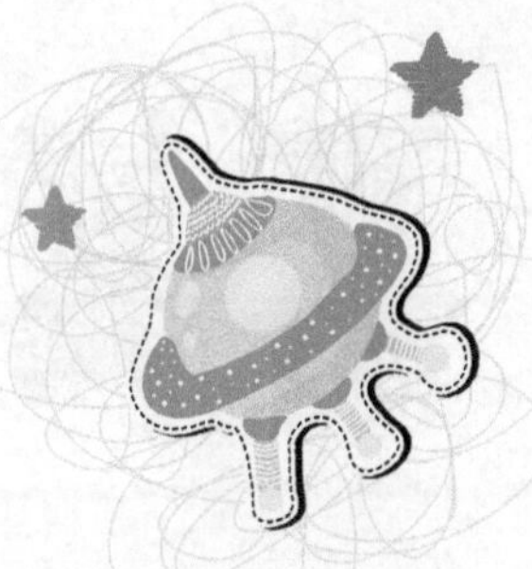

Today you will practice making nouns plural by adding the correct ending.

Directions: Add the correct ending to each word (-s or -es) to make the noun plural.

1. Kiss ☐
2. Chair ☐
3. Bush ☐
4. Rug ☐
5. Brush ☐
6. House ☐
7. Doctor ☐
8. Kid ☐
9. Store ☐
10. Lunch ☐

FITNESS PLANET

Let's get some fitness in! Go to page 111 to try some fitness activities.

Cross out objects to solve the subtraction problem.

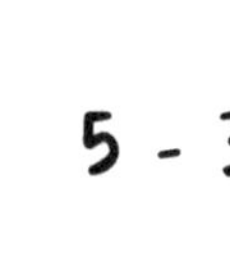

1. 2 - 2 = ☐

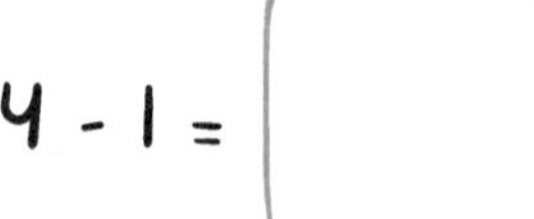

2. 5 - 3 = ☐

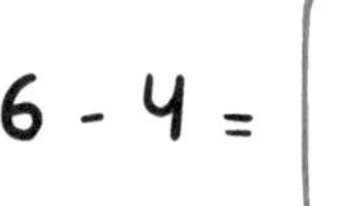

3. 4 - 1 = ☐

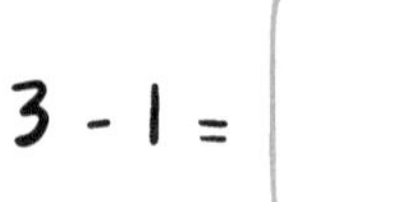

4. 6 - 4 = ☐

5. 3 - 1 = ☐

6. 8 - 5 = ☐

Topic 1 Sequencing

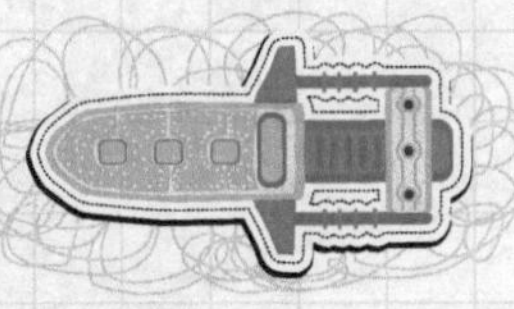

Today, you will practice sequencing, or putting events in order.

Directions: Use the transition words provided below to draw pictures of how you get ready for school each day. Be sure to draw the events in order from first to last.

First	
Next	
Then	
After that	
Last	

Kindergarten

WEEK 4

Now it's time for:

- number bonds
- letters Gg and Hh
- beginning sounds
- addition word problems and more!

Topic 1 Number Bonds to 10

Complete the ten frame to find the missing number to make 10.

1. 7 + ◯ = 10  

2. 5 + ◯ = 10  

3. 2 + ◯ = 10  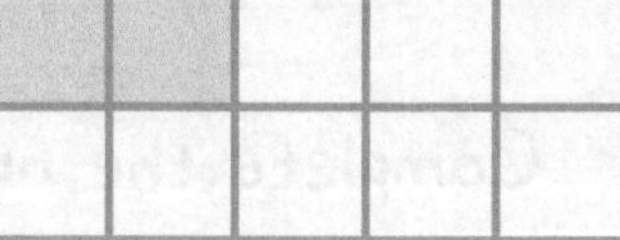

4. ◯ + ◯ = 10  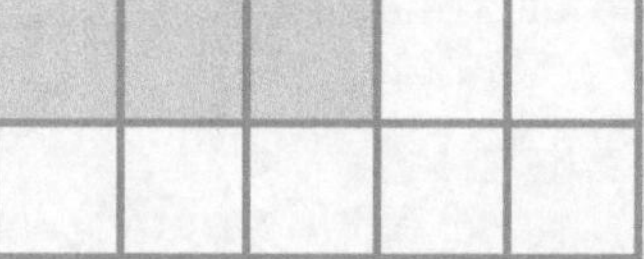

5. 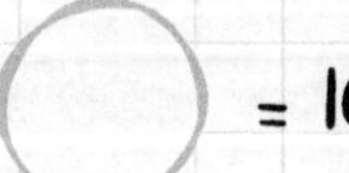 + ◯ = 10  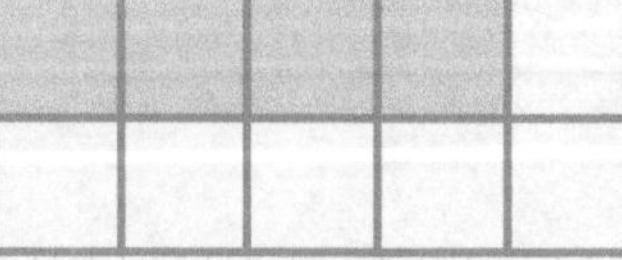

6. ◯ + ◯ = 10 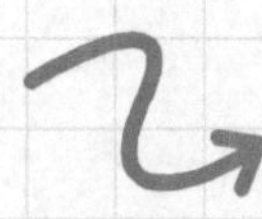 

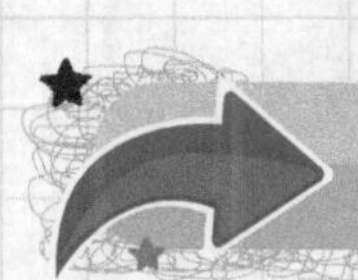

FITNESS PLANET  Let's get some fitness in! Go to page 111 to try some fitness activities.

1. Complete the number bond.

10 — 5, ☐

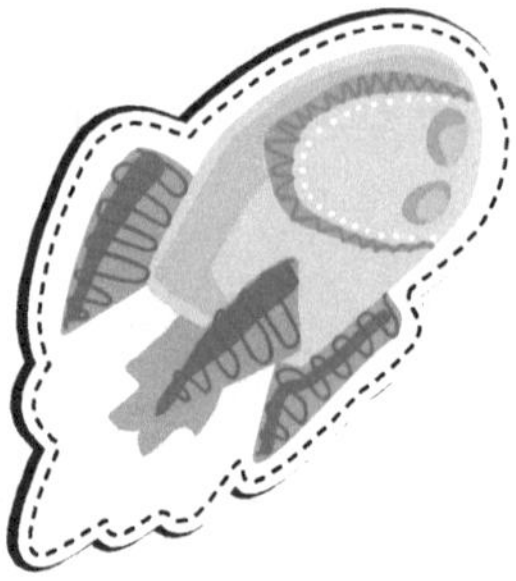

2. Complete the number bond.

6 — 4, ☐

3. Complete the number bond.

☐ — 5, 5

4. Complete the number bond.

☐ — 3, 3

5. Complete the number bond.

9 — 5, ☐

6. Complete the number bond.

4 — 4, ☐

Topic 1 Letters Gg and Hh

Today you are going to review and practice printing the letters Gg and Hh.

Remember Gg is for gum and Hh is for hat.

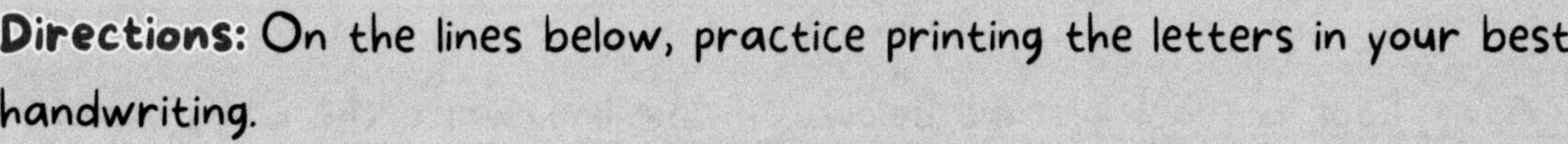

Directions: On the lines below, practice printing the letters in your best handwriting.

G ....

g ....

H ....

h ....

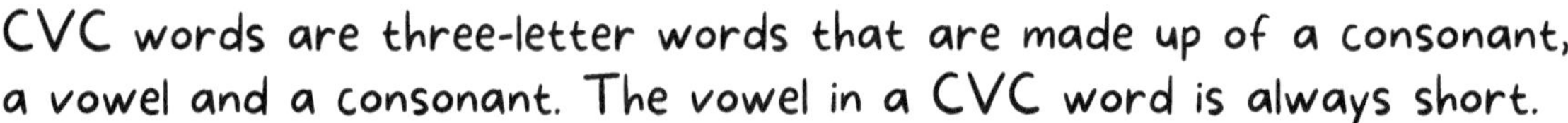

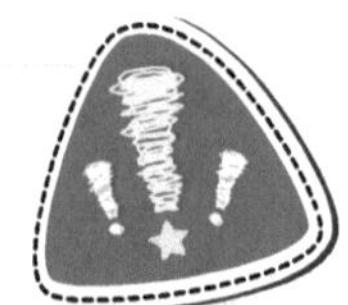

CVC words are three-letter words that are made up of a consonant, a vowel and a consonant. The vowel in a CVC word is always short.

Today, you are going to practice identifying the beginning sounds in words and writing the corresponding letter.

Directions: Look at each picture below. On the line, write the CVC word that matches the picture and identify the beginning sound by circling it.

1. ....

2. ....

3. ....

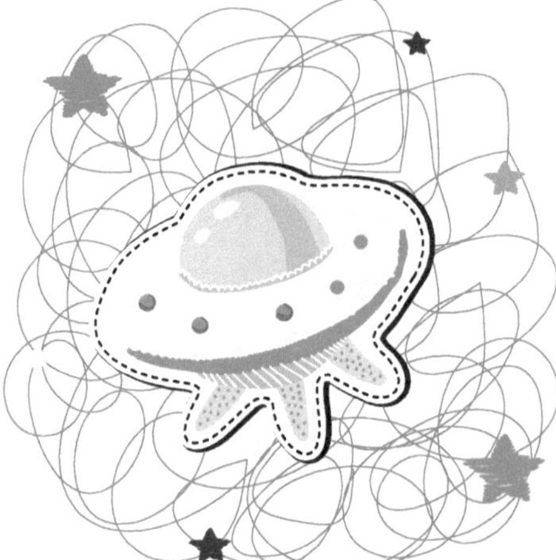

4. ....

5. ....

Topic 1 Pushing and Distance

Last week, you explored the forces of pushing and pulling by brainstorming objects that move in these ways.

Today, you will explore how the distance an object moves depends on how hard it is being pushed.

Materials

* Large building blocks
* A piece of wood or wooden train track
* Small toy car
* Tape measure

Procedure

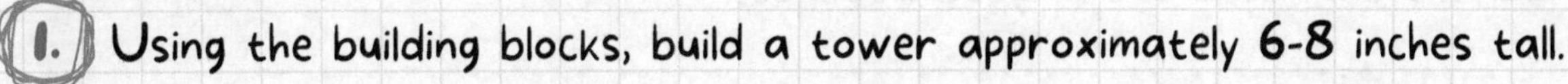

1. Using the building blocks, build a tower approximately **6-8** inches tall.

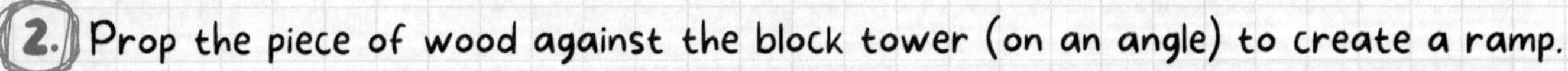

2. Prop the piece of wood against the block tower (on an angle) to create a ramp.
3. Push the toy car down the ramp several times with different amounts of force.
4. Measure how far the car moves each time. Record your answers.

5. Answer the questions that follow.

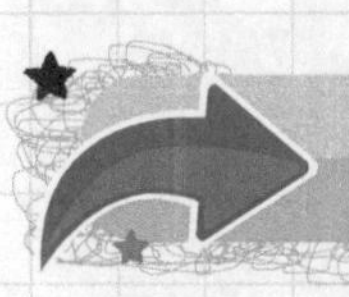

FITNESS PLANET

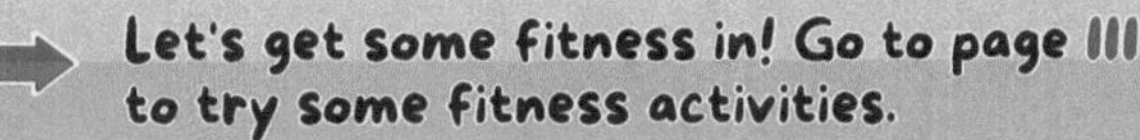

Let's get some fitness in! Go to page 111 to try some fitness activities.

Amount of Force	Distance Traveled
Light	
Medium	
Hard	

Follow-Up Questions

1. When did the car move the shortest distance?

....

....

....

2. When did the car move the longest distance?

....

....

....

3. Why?

....

....

....

1. Write the addition problem that the number bond shows.

☐ + ☐ = ☐

3 — 1, 2

2. Write the addition problem that the number bond shows.

☐ + ☐ = ☐

10 — 3, 7

3. Write the addition problem that the number bond shows.

☐ + ☐ = ☐

9 — 3, 6

4. Write the subtraction problem that the number bond shows.

☐ - ☐ = ☐

8 — 4, 4

5. Write the subtraction problem that the number bond shows.

☐ - ☐ = ☐

6 — 3, 3

6. Write the subtraction problem that the number bond shows.

2 — 1, 1

Week 4 Operations & Algebraic Thinking

Topic 4 Addition Word Problems

1. Use the images to solve the word problem. 1 ladybug was sitting on a log. 3 more ladybugs came. How many ladybugs are there in all?

2. Use the images to solve the word problem. Jim sees 3 stars. Then he sees 2 more. How many stars does Jim see in all?

3. Use the ten frame to solve the word problem. There are 3 girls and 1 boy. How many people are there in all?

4. Use the ten frame to solve the word problem. I have three pencils and 3 rules. How many things do I have in all?

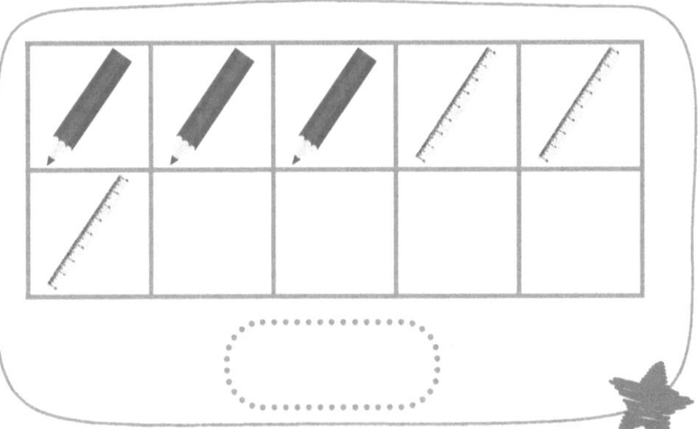

5. Use the number line to solve the word problem. There are 2 ducks. 2 more ducks join them. How many ducks are there in all?

6. Use the number line to solve the word problem. I see 3 bugs. 5 more bugs come. How many bugs are there in all?

It is important to be able to retell a story after you have listened to or read it. To **retell** a story means to tell someone what happened, or what it was about, in your own words. When retelling a story, it is important to tell about the events in sequence, or the order in which they happened.

Today, you are going to practice retelling a story with a family member at home. You will need to choose a book from your house or the library.

Directions: Choose a book to read to someone in your family. Read the story aloud to them. Then, use the questions below to help you retell the story aloud to them. You do not have to write out answers to these questions.

1. What was the title of the story? ....

2. Who are the characters in the story? ....

3. What was the setting of the story? ....

4. What happened in the story? ....

5. How did the story end? ....

6. Did you like this story? Why or why not? ....

Let's get some fitness in! Go to page 111 to try some fitness activities.

Topic 4 Question Words

Question words are the words used to begin sentences that are questions. There are many possible question words, but the most common are:

1. Who
2. What
3. Why
4. When
5. Where
6. How

Today, you will practice asking questions using the words above by interviewing someone in your family. Interview means to ask someone questions about themselves so you can learn more about them.

Directions: Choose a person to interview. Use each question word given below to write a question that you would like to ask that person.

Examples:

What is your favorite book? Where do you like to eat?

1. Who ....
2. What ....
3. Why ....
4. When ....
5. Where ....
6. How ....

Now use your questions to interview the person you chose!

Topic 5 Subtraction Word Problems

1. Use the images to solve the word problem. There were 6 butterflies. 2 flew away. How many are left?

2. Use the images to solve the word problem. There were 5 cars. 1 drove away. How many are left?

3. Use the ten frame to solve the word problem. There are 9 dogs. 3 run away. How many are left?

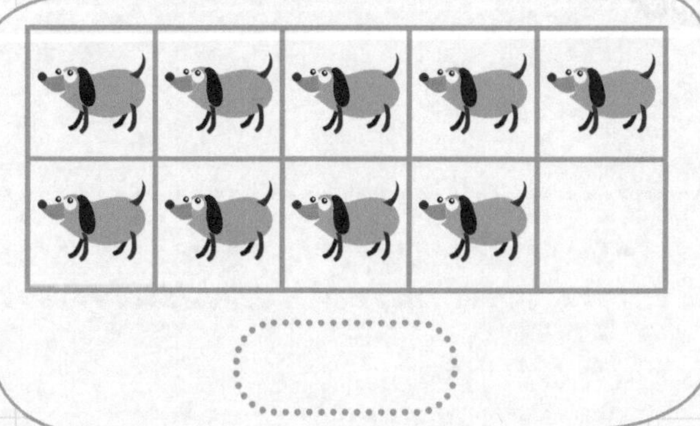

4. Use the ten frame to solve the word problem. There are 7 cats. 4 run away. How many are left?

5. Use the number line to solve the word problem. I see 6 bats. 3 bats fly away. How many are left?

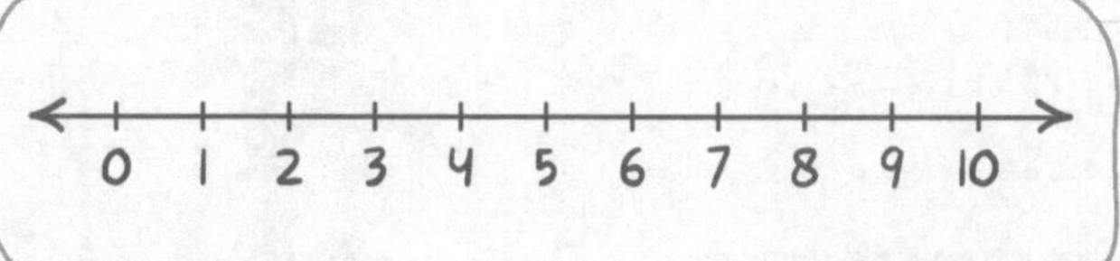

6. Use the number line to solve the word problem. I see 10 bugs. 3 bugs crawl away. How many are left?

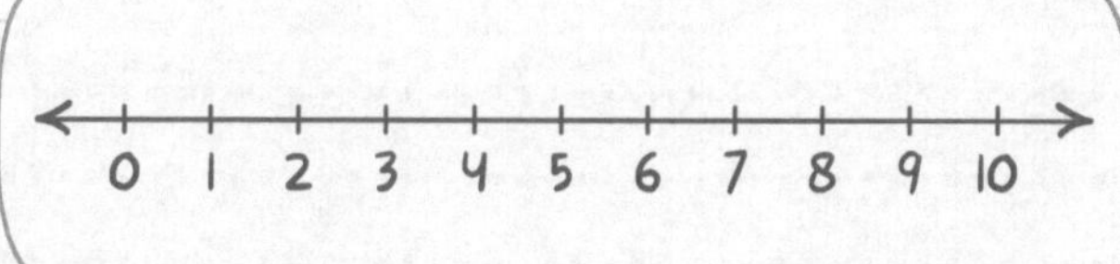

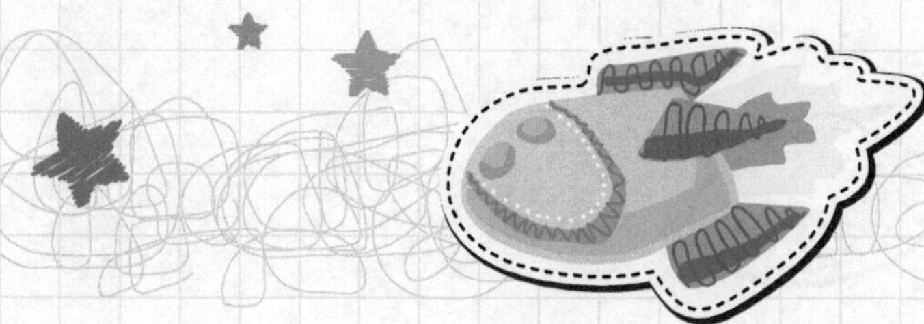

A **community helper** is a person who is helpful to others in places like schools, neighborhoods and cities.

Examples of Community Helpers:

1. parents
2. police officers
3. firefighters
4. teachers

Directions: Listed below are 3 different community helpers. Write a sentence or two to describe how each person is a helper.

1. School Bus Driver

2. Nurse

3. Police officer

Kindergarten

WEEK 5

Let's see how well you know:

- place value
- letters Ii and Jj
- middle sounds
- fiction vs. nonfiction and more!

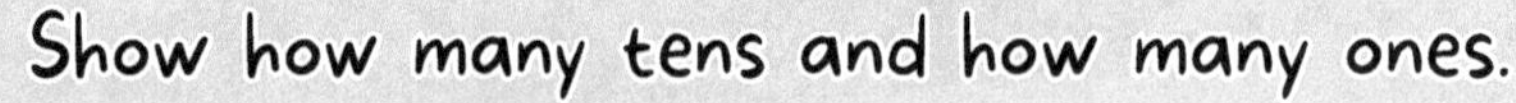

Show how many tens and how many ones.

1. ___ tens ___ ones

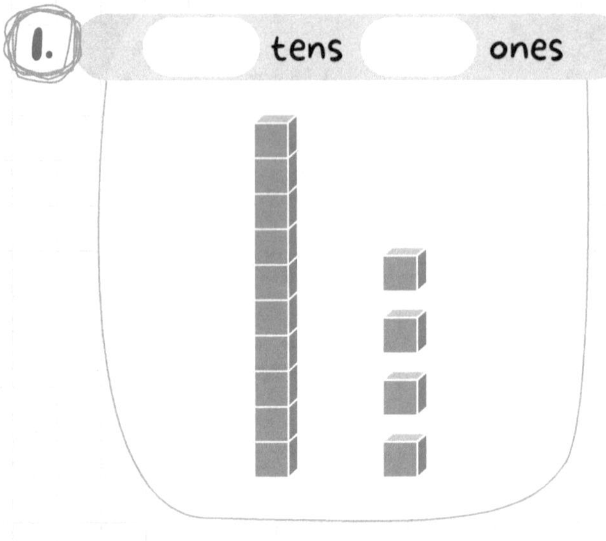

2. ___ tens ___ ones

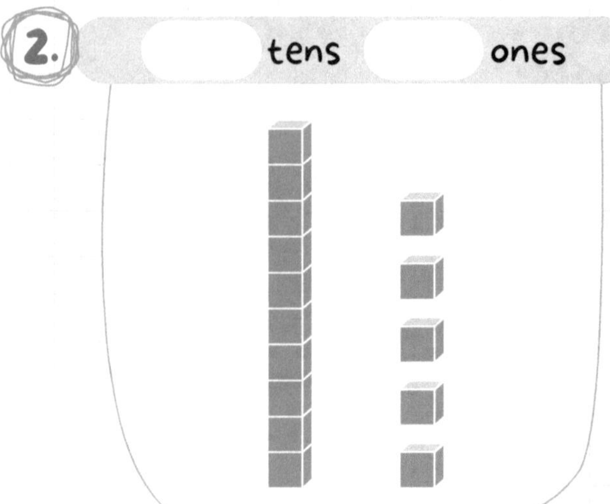

3. ___ tens ___ ones

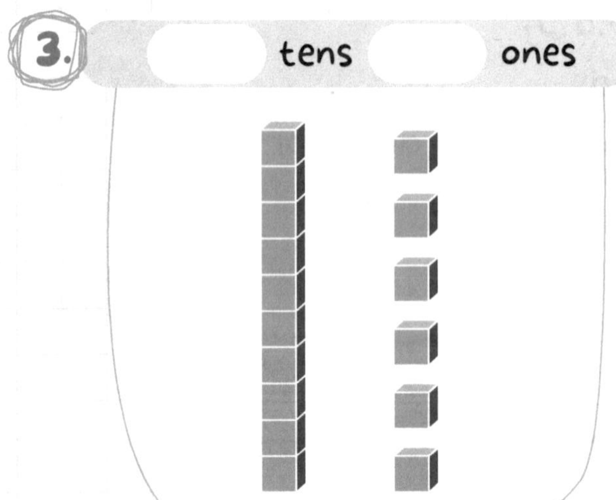

4. ___ tens ___ ones

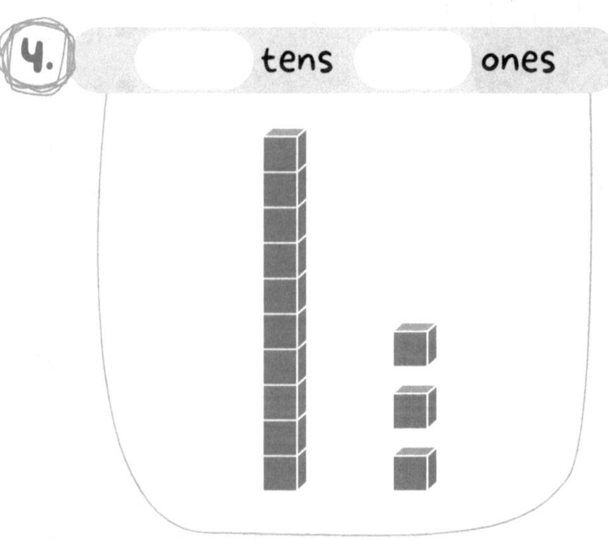

5. ___ tens ___ ones

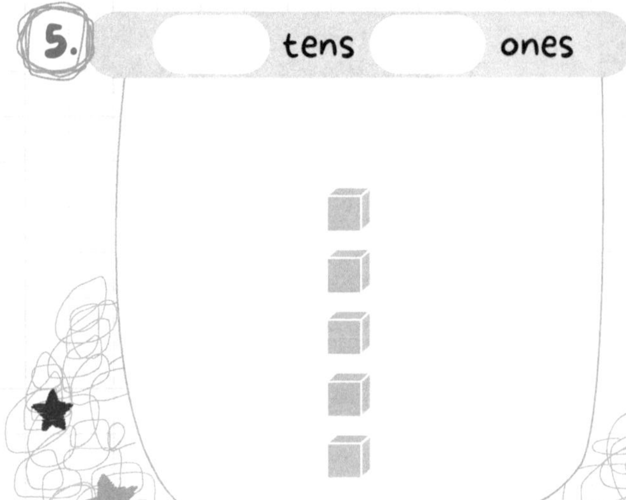

6. ___ tens ___ ones

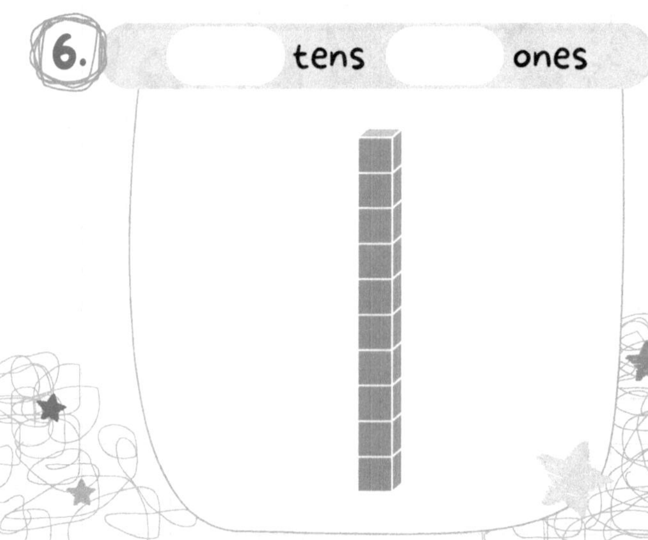

1. 1 tens, 4 ones is the same as

2. 1 tens, 7 ones is the same as

3. 0 tens, 4 ones is the same as

4. 0 tens, 9 ones is the same as

5. 1 tens, 0 ones is the same as

6. 1 tens, 6 ones is the same as

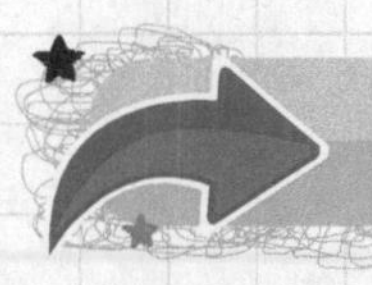

FITNESS PLANET
Let's get some fitness in! Go to page 111 to try some fitness activities.

Week 5 Handwriting Review

Topic 1 Letters Ii and Jj

Today you are going to review and practice printing the letters Ii and Jj.

Remember Ii is for igloo and Jj is for jar.

Directions: On the lines below, practice printing the letters in your best handwriting.

I ....

i ....

J ....

j ....

CVC words are three-letter words that are made up of a consonant, a vowel and a consonant. The vowel in a CVC word is always short.

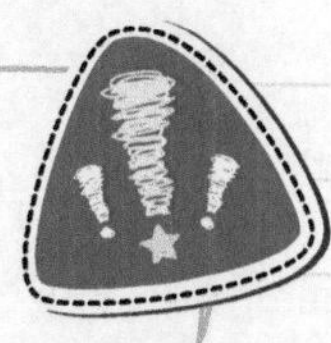

Today, you are going to practice identifying the middle sounds in words and writing the letter that makes that sound.

Directions: Look at each picture below. On the line, write the CVC word that matches the picture and identify the middle sound by circling it.

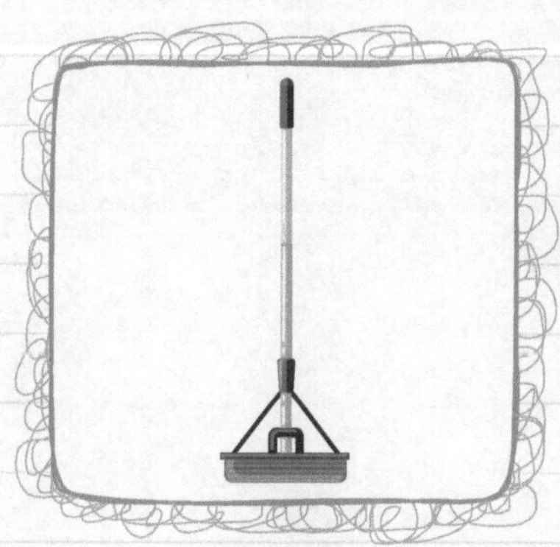

1. ....

2. ....

3. ....

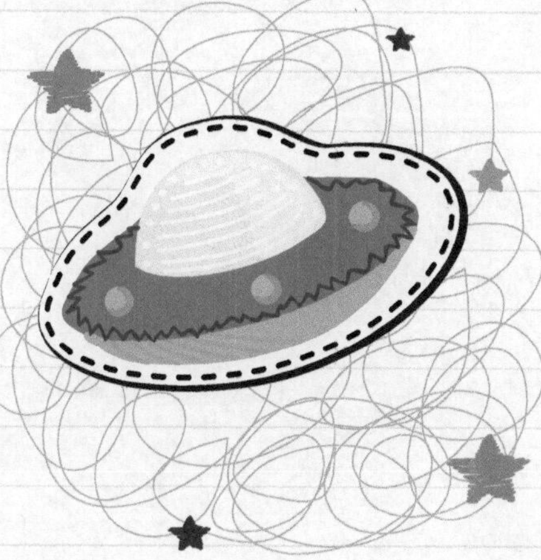

4. ....

5. ....

Topic 1 Day Sky vs. Night Sky

Today, you will explore the differences between the objects observed in the day sky and the night sky.

Directions: Observe the sky during the day and at night. Draw pictures of your observations below. Then, answer the questions that follow.

Day Sky:

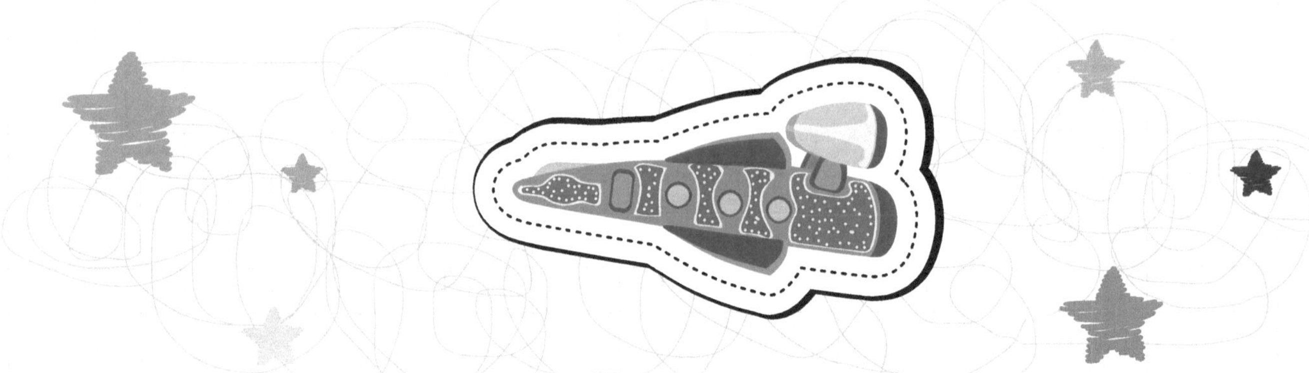

Topic 1 Day Sky vs. Night Sky

Night Sky:

Follow-Up Questions:

1. What do you see in the day sky?

....

....

....

2. What do you see in the night sky?

....

....

....

3. What else is different about the day sky and night sky?

....

....

....

FITNESS PLANET

Let's get some fitness in! Go to page 111 to try some fitness activities.

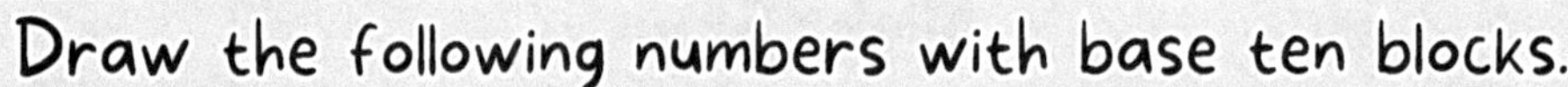

Draw the following numbers with base ten blocks.

1. 12

2. 6

3. 11

4. 10

5. 17

6. 8

Complete the problem to make the addition sentence true.

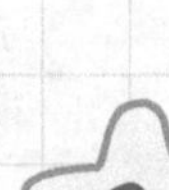

1. 10 + ☐ = 16

2. 10 + ☐ = 11

3. 10 + ☐ = 10

4. ☐ + 4 = 14

5. ☐ + 8 = 8

6. ☐ + 2 = 12

Topic 3 Fiction vs. Nonfiction

In the past several lessons, you have been reviewing elements of fictional stories. Today, you are going to learn about the difference between fiction and nonfiction.

Fiction is when books are about stories that are made up. The characters and events in the story are not real.

Important elements of fiction include:

- Made up characters, settings and events
- Illustrations that are drawn or created by an artist
- Animals may be the main characters
- Animals or objects may talk
- The story is just for fun

Nonfiction is when books are about real people or events. They give information or teach the reader about something.

Important elements of nonfiction include:

- A real topic (for example, the book is about a real person, place or thing)
- Pictures are often real photographs
- The book is teaching the reader about a certain topic

Directions: Answer the following questions.

1. Fiction is:

   A. Real   **B.** Made up

2. Nonfiction is:

   A. Real   **B.** Made up

3. Give 2 examples of books you have read that are fiction.

A. ....

B. ....

4. How did you know these books were fiction? ....

....

5. Give 2 examples of books you have read that are nonfiction.

A. ....

B. ....

6. How did you know these books were nonfiction? ....

....

7. Which type of book often uses photographs as the pictures in the book?

A. Fiction B. Nonfiction

8. Which type of book may have animals that talk as characters in the story?

A. Fiction B. Nonfiction

9. Do you like fiction or nonfiction better? Why? ....

....

FITNESS PLANET → Let's get some fitness in! Go to page 111 to try some fitness activities.

Topic 3 Measurement

Use the ruler to measure to the nearest inch.

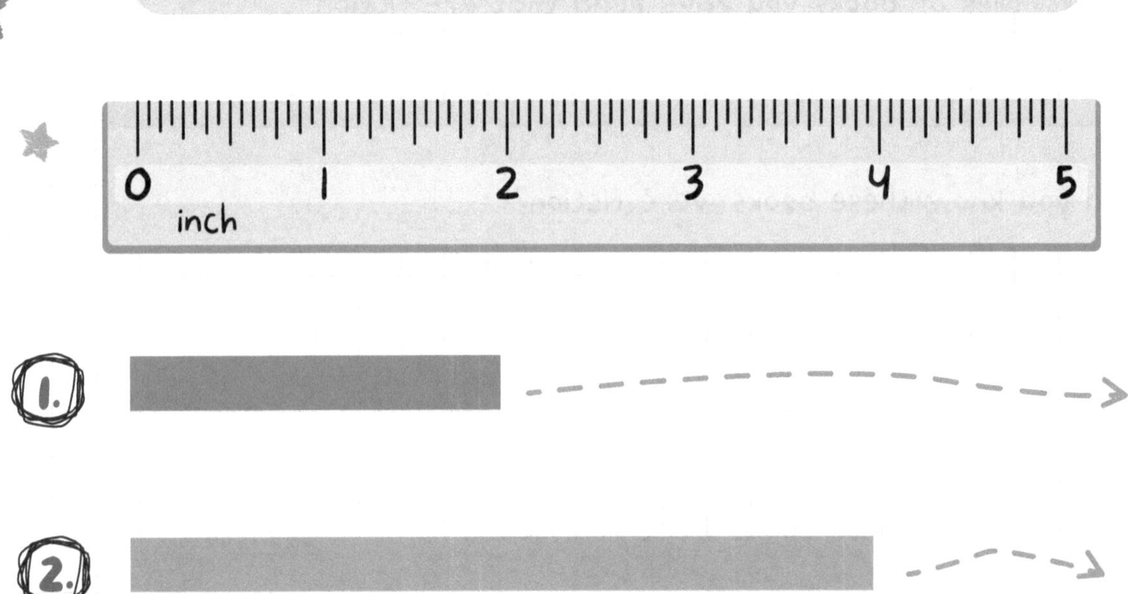

1.

2.

3.

4.

5.

6.

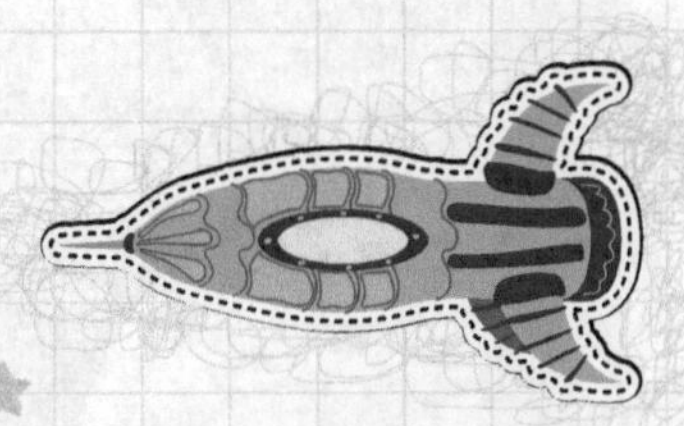

Rules are important because they help to maintain order and keep people safe.

Directions: List 3 rules from your classroom or at home and describe why they are important.

1. Rule #1:

   It's important because:

2. Rule #2:

   It's important because:

3. Rule #3:

   It's important because:

Kindergarten

WEEK 6

It's time to explore:

- measurement
- letters Kk and Ll
- ending sounds
- weather and more!

How long is the object?

1. ☐ blocks long

2. ☐ blocks long

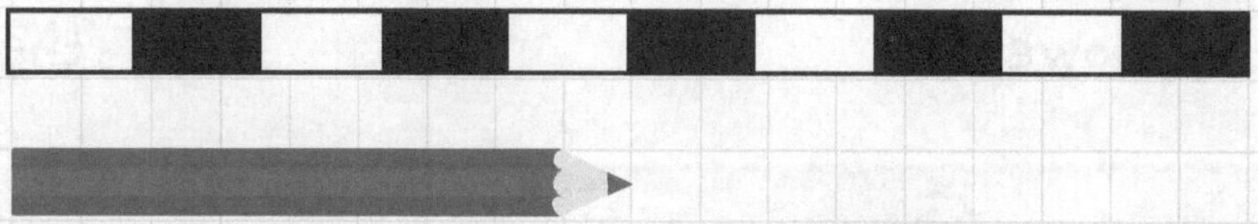

3. ☐ blocks long

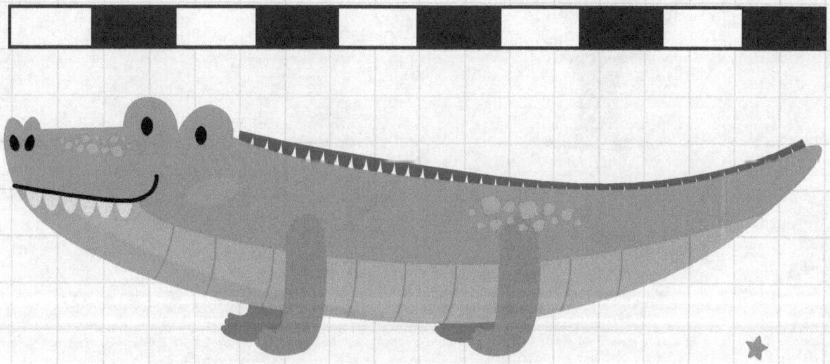

4. ☐ blocks long

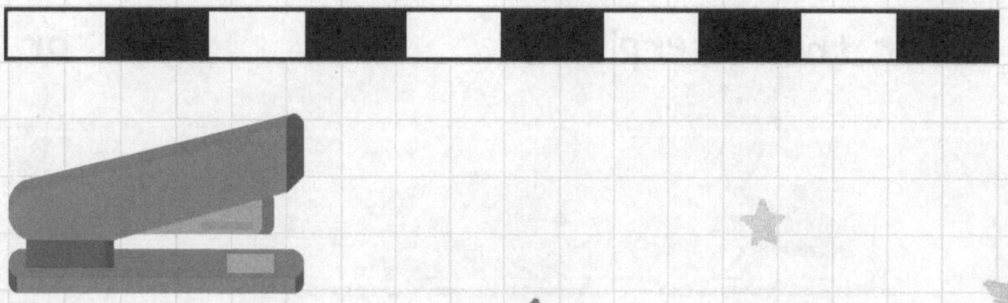

FITNESS PLANET
Let's get some fitness in! Go to page 111 to try some fitness activities.

1. Which is longer, red or blue?

2. Which is longer, red or blue?

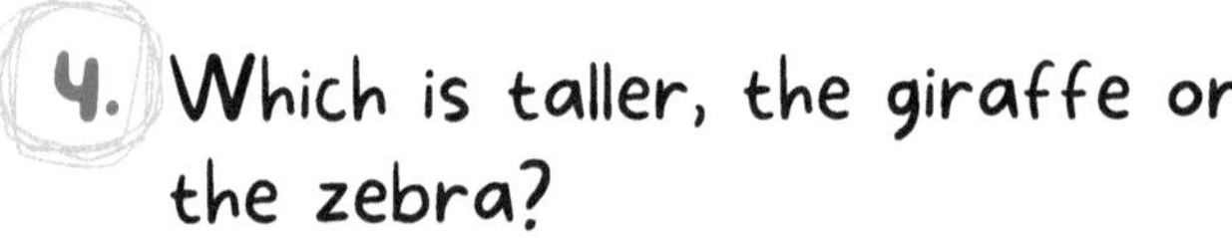

3. Which is taller, the tree or the flowers?

4. Which is taller, the giraffe or the zebra?

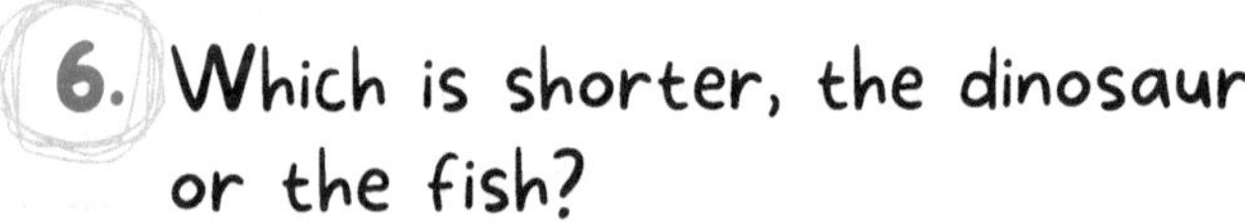

5. Which is shorter, the cricket or the caterpillar?

6. Which is shorter, the dinosaur or the fish?

Week 6 Handwriting Review

Topic 1 Letters Kk and Ll

Today you are going to review and practice printing the letters Kk and Ll.
Remember **Kk** is for kangaroo and **Ll** is for lemon.

Directions: On the lines below, practice printing the letters in your best handwriting.

K ....

k ....

L ....

l ....

CVC words are three-letter words that are made up of a consonant, a vowel and a consonant. The vowel in a CVC word is always short.

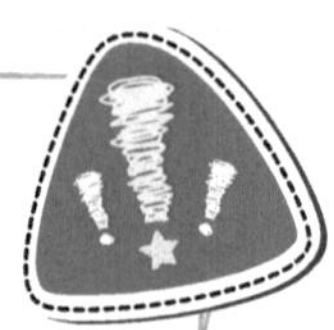

Today, you are going to practice identifying the ending sounds in words and writing the corresponding letter.

Directions: Look at each picture below. On the line, write the CVC word that matches the picture and identify the ending sound by circling it.

1. ....

2. ....

3. ....

4. ....

5. ....

Today, you will observe the weather in your city and determine how it changes over time.

Directions: Observe the weather each day for one week. Record your observations below. Then, answer the questions that follow.

Day	Weather
1	
2	
3	
4	
5	
6	
7	

FITNESS PLANET → Let's get some fitness in! Go to page 111 to try some fitness activities.

Follow-Up Questions:

1. How was the weather the SAME during the week?

2. How was the weather DIFFERENT during the week?

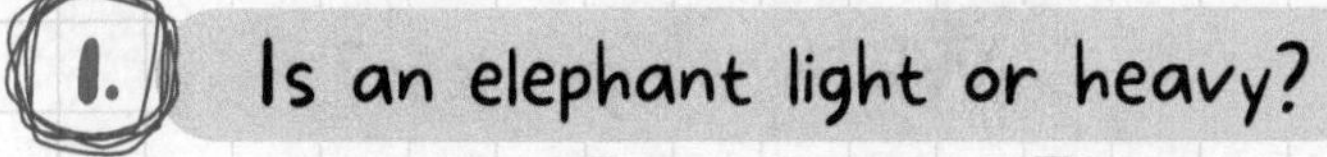

1. Is an elephant light or heavy?

2. Is a mouse light or heavy?

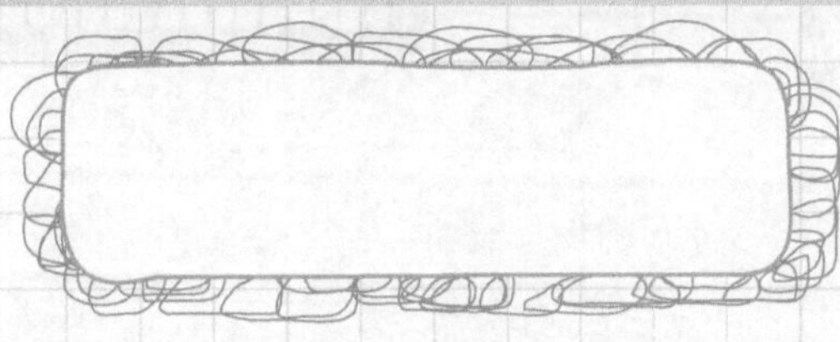

3. Is a feather light or heavy?

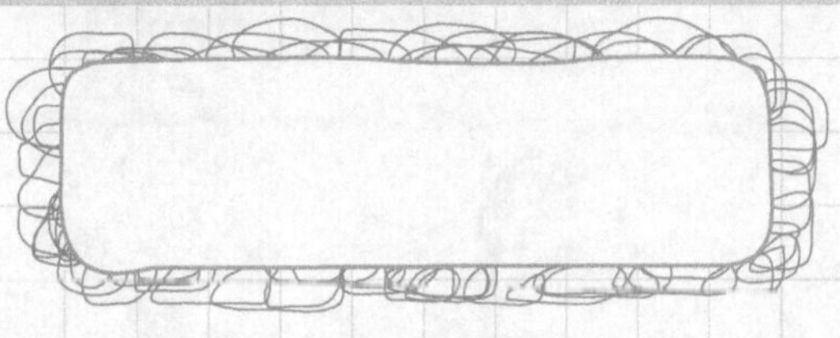

4. Is a tractor light or heavy?

5. Is a penny light or heavy?

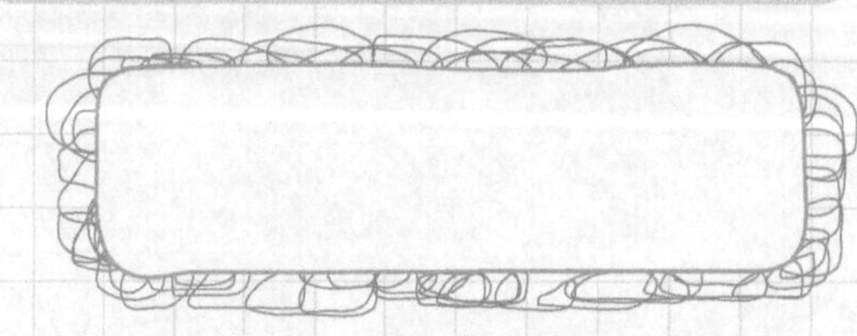

6. Is a fish tank light or heavy?

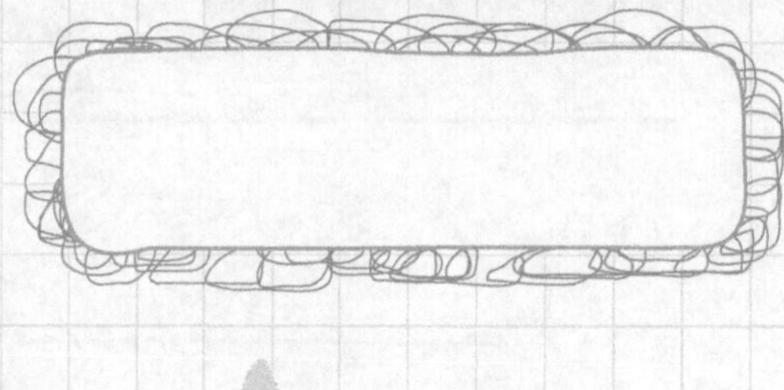

Which weighs more?

1.

2.

3.

Which weighs less?

4.

5.

6.

Topic 3 Informational Text

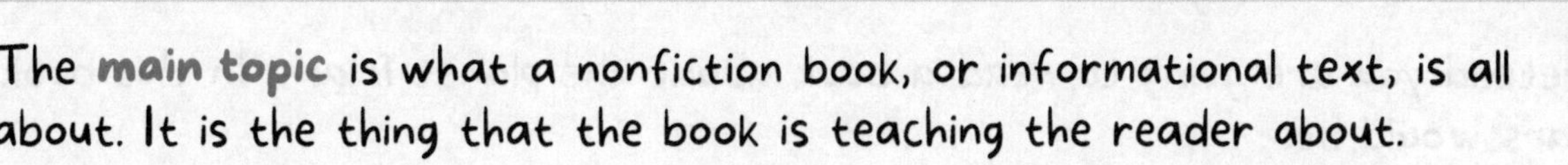

The **main topic** is what a nonfiction book, or informational text, is all about. It is the thing that the book is teaching the reader about.

Examples of topics might be:

* A specific animal
* A famous person
* A city, state or country
* Planets
* Vehicles

Key details are all the information in the book that the author uses to teach the reader about the topic.

Examples of key details an author might include in a book about sloths might include:

* What sloths look like
* What sloths eat
* Where sloths live
* How slowly sloths move
* What sloths like to do

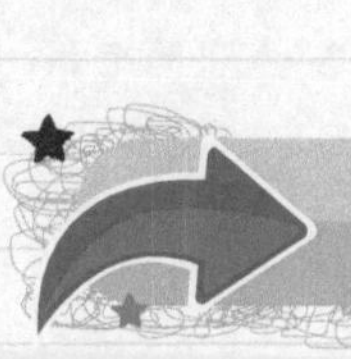

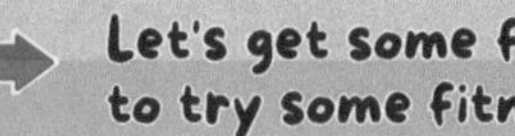

Let's get some fitness in! Go to page 111 to try some fitness activities.

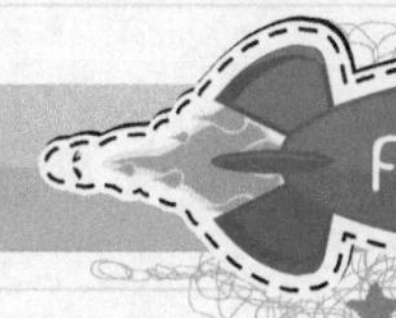

Directions: Answer the questions below.

1. Pretend you are going to read a book about the planet Mars. In this book, Mars would be:

   A. The topic

   B. A detail

2. List **2** key details that might be included in a book about Mars.

   A. ....

   B. ....

3. What might the topic of a book be, if a key detail was "Where the President Lives?"

   ....

4. What might the topic of a book be, if a key detail was "What Does a firefighter Do?"

   ....

Directions: Pretend you are writing a nonfiction book. Answer the questions below.

1. What would the main topic of your book be? ....

   ....

2. List **3** key details you would include in your book.

   A. ....

   B. ....

   C. ....

3. What is the purpose of your book? ....

   ....

Which object holds more?

1.

2.

3.

4.

5.

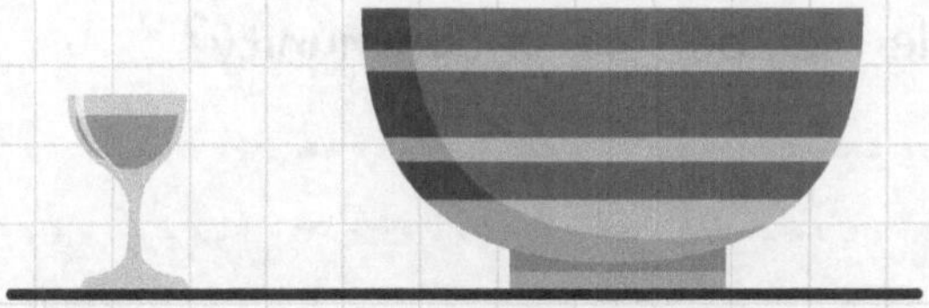

6.

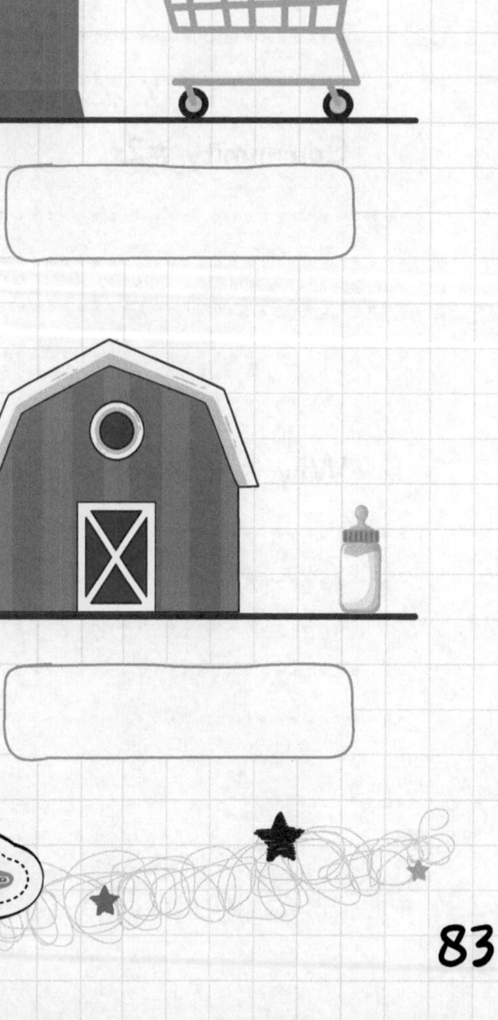

You are a member of many different communities. Some examples might include your: family, school, neighborhood, church and city. It is important to be a responsible member of each community you're a part of.

Directions: On the lines below, list **2** different communities that you are part of. Then, write a sentence or two to describe how you can be a responsible member of that community.

1. Community #1:

2. Community #2:

3. Why is it important to be a responsible member of a community?

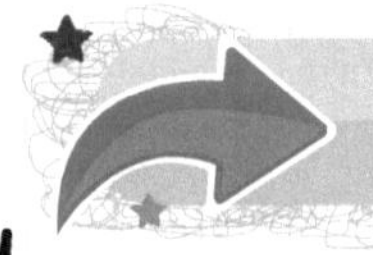

FITNESS PLANET

Let's get some fitness in! Go to page 111 to try some fitness activities.

Kindergarten

WEEK 7

Let's have fun with:

- substituting sounds
- the environment
- 2D shapes
- spatial terms and more!

Look at the picture and answer the following questions.

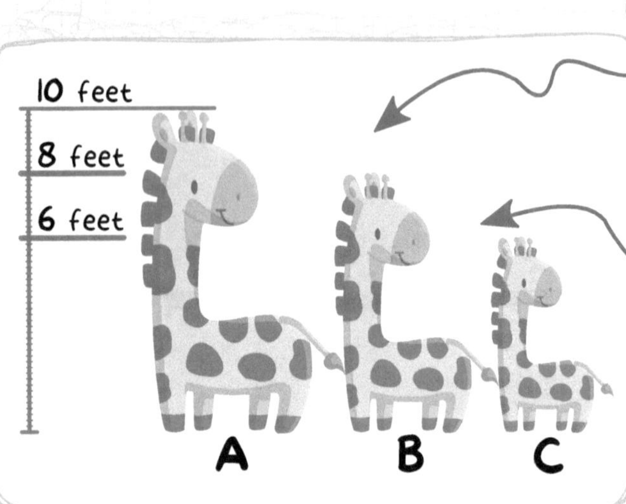

1. How tall is giraffe A?

....

....

2. How tall is giraffe B?

....

....

3. How tall is giraffe C?

....

....

4. How tall is ladder A?

....

....

5. How tall is ladder B?

....

....

6. How tall is ladder C?

....

....

6 feet
4 feet
2 feet
A B C

1. Are there more sharks or elephants?

2. Are there more monkeys or crabs?

3. Are there less sharks or elephants?

4. Are there less monkeys or crabs?

5. Are there more rhinos or roosters?

6. Are there less rhinos or roosters?

Week 7 Handwriting Review

Topic 1 Letters Mm, Nn and Oo

Today you are going to review and practice printing the letters Mm, Nn and Oo. Remember **Mm** is for mitten, **Nn** is for net and **Oo** is for octopus.

Directions: On the lines below, practice printing the letters in your best handwriting.

M ....

m ....

N ....

n ....

O ....

o ....

Week 7 Phonics Review

Topic 2 Substituting Sounds

In many CVC (consonant-vowel-consonant) words, you can substitute, or change, the beginning, middle or ending sound to make a new word. For instance, in the word shown below, if the beginning sound is changed from /m/ to /r/, the word becomes ran.

man → **ran**

Today, you are going to practice changing the sounds in words to make new words.

Directions: In the words below, change the beginning sound to make a new word. Write the new word on the line.

1. mug → ....
2. rob → ....
3. sad → ....

Directions: In the words below, change the middle sound to make a new word. Write the new word on the line.

4. big → ....
5. let → ....
6. rat → ....

Directions: In the words below, change the ending sound to make a new word. Write the new word on the line.

7. sit → ....
8. bet → ....
9. pop → ....

FITNESS PLANET → Let's get some fitness in! Go to page 111 to try some fitness activities.

FITNESS

Topic 1 Helping the Environment

Humans can make big changes, both good and bad, in our environment. The environment is the land, air and water around us.

Today, you are going to think of four ways you could help make your environment better.

Directions: Think of 4 ways you can make the environment better or cleaner. Draw a picture for each idea below.

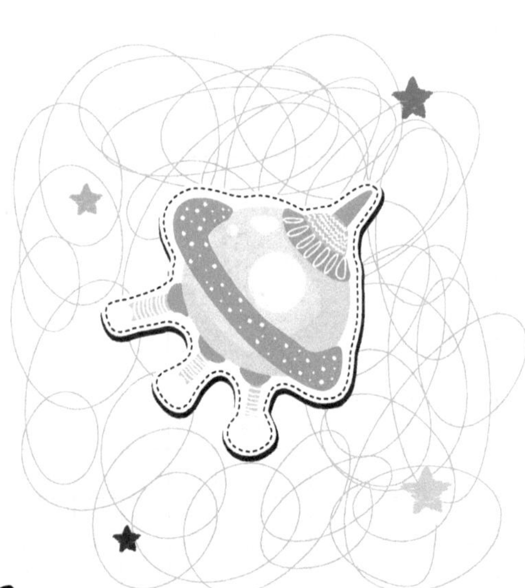

Topic 1 Helping the Environment

1. Circle the carrot that is longer.

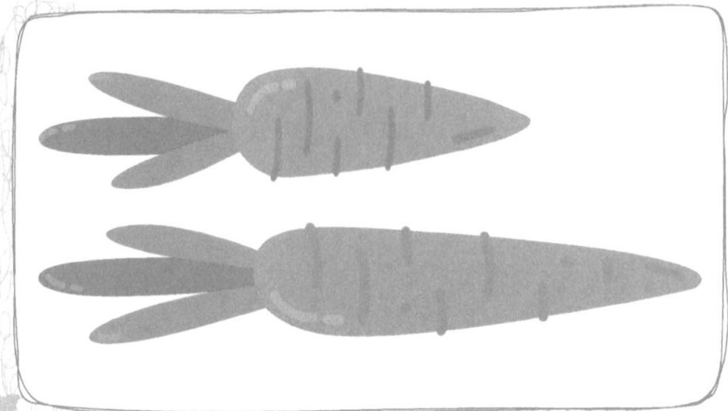

2. Circle the animal that is shorter.

3. Circle the sheep that is bigger.

4. Circle the person that is younger.

5. Circle the thing that is heavier.

6. Circle the flower that is shorter.

Name the 2D shape.

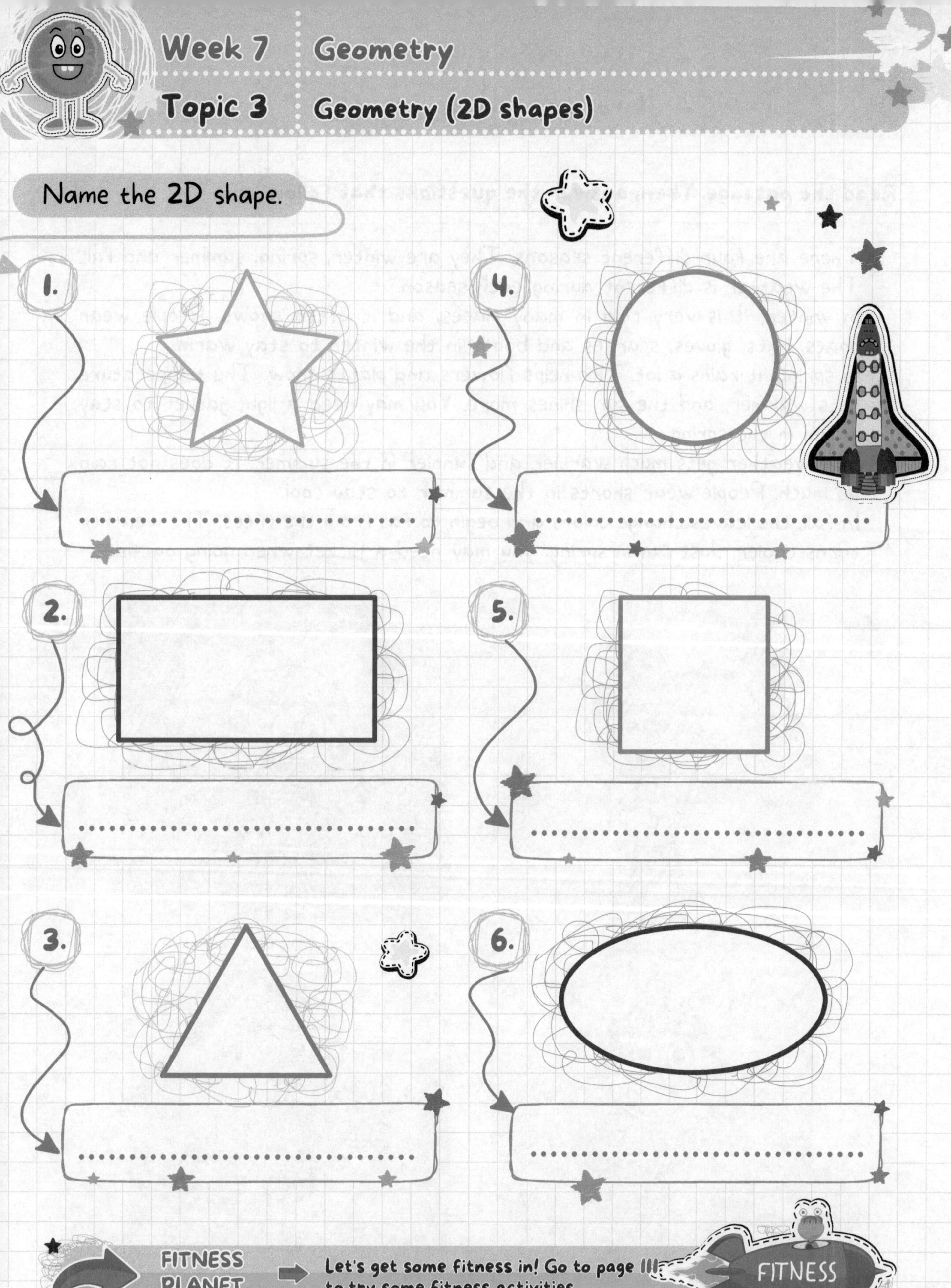

1. ....

2. ....

3. ....

4. ....

5. ....

6. ....

FITNESS PLANET

Let's get some fitness in! Go to page 111 to try some fitness activities.

FITNESS

Read the passage. Then, answer the questions that follow.

There are four different seasons. They are winter, spring, summer and fall. The weather is different during each season.

In winter, it is very cold in many places, and it often snows. People wear coats, hats, gloves, scarves and boots in the winter to stay warm.

In spring, it rains a lot. This helps flowers and plants grow. The temperature gets warmer, and the sun shines more. You may need a light jacket to stay warm in the spring.

The weather gets much warmer and sunnier in the summer. It does not rain as much. People wear shorts in the summer to stay cool.

In fall, the leaves change colors and begin to fall from the trees. The weather turns cooler. Just like in spring, you may need a jacket when going outside.

Now answer the questions about the passage.

1. What is the main topic of this passage?

2. List 3 key details given in the passage.

3. What do you think the word temperature means in the passage?

4. What type of clothes do people wear in the winter?

5. In which season might you need to wear a coat?

6. When do the leaves begin to change colors?

7. What is the weather like in spring?

8. What is your favorite season? Why?

9. Draw a picture of an activity you like to do during your favorite season.

How many sides does each shape have?

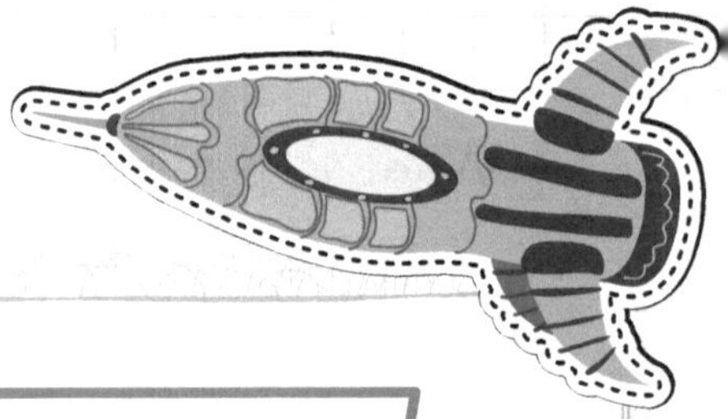

1.

....

4.

....

2.

....

5.

....

3.

....

6.

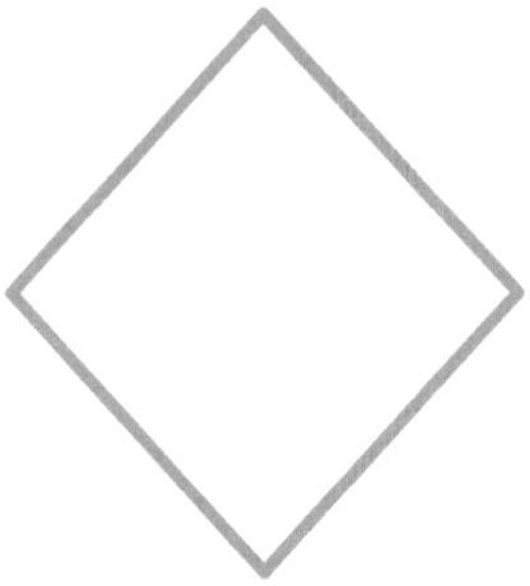

....

Topic 1 Spatial Terms

Spatial terms are words that are related to location, direction and distance. See the table below for examples.

here	over there	left	right
forward	under	between	above

Today, you are going to practice using spatial terms by giving directions to a specific location.

Directions: Choose a location in your house to give directions to. Ask a family member to follow all the directions you give. Use spatial terms such as those listed above to help your family member get to the correct location in your house. See the example below to help guide you.

Examples

Walk forward 10 steps. Turn left and walk 10 more steps. Turn right and walk 3 steps between the wall and the table. Turn left. Walk 5 steps under the archway. Reach out and turn the doorknob to open the door. Walk through the door and turn left. You should be in the pantry.

Now it's your turn. Practice giving directions to a family member. Then, answer the questions below.

1. Did your family member end up in the correct location? Why or why not?

....

....

....

2. What do you need to change about your directions to help them get to the correct place?

....

....

....

Let's get some fitness in! Go to page 111 to try some fitness activities.

Kindergarten

WEEK 8

Now try your best at:

- letters Pp and Qq
- sound fluency
- poetry
- addresses and more!

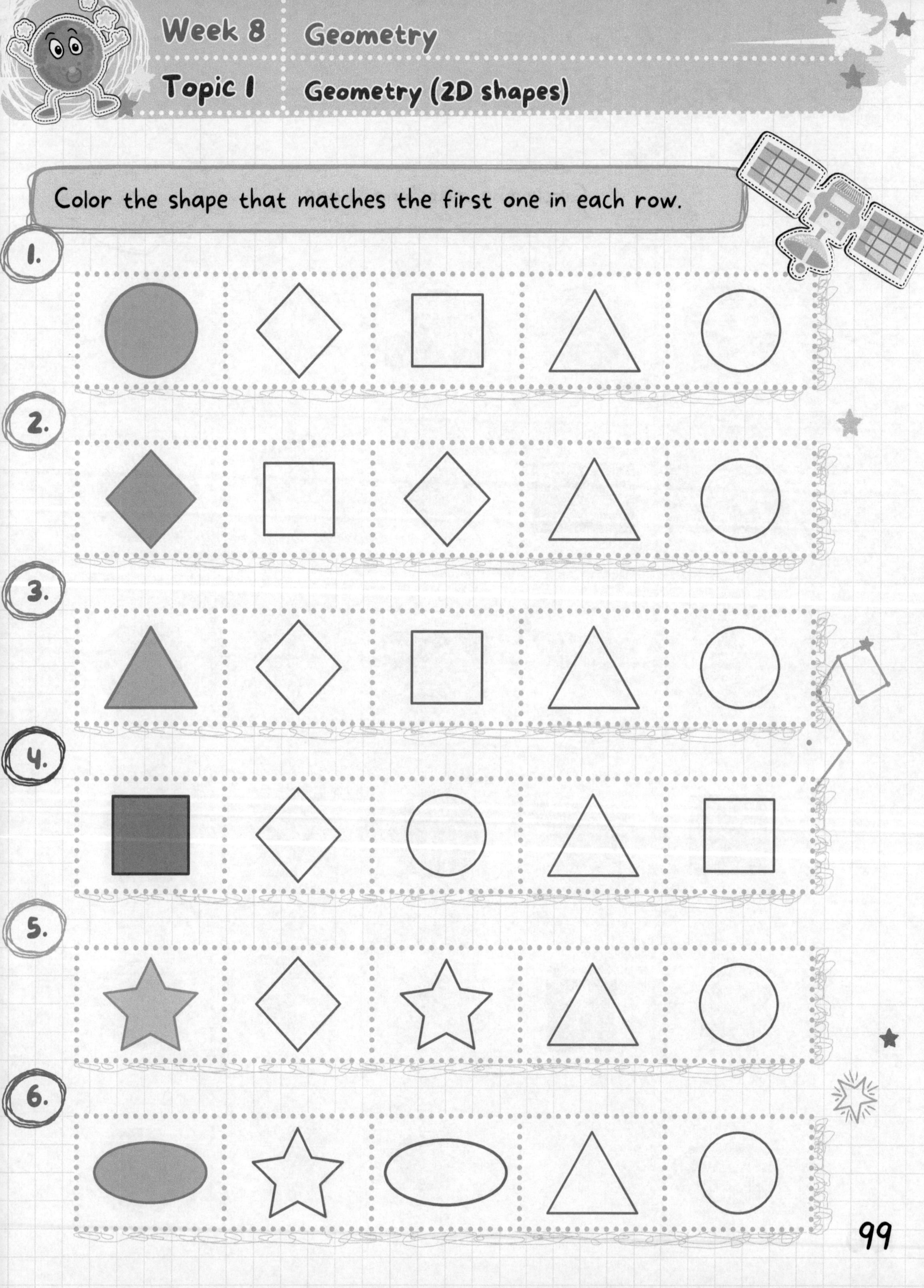
Week 8 Geometry
Topic 1 Geometry (2D shapes)
Color the shape that matches the first one in each row.
1.
2.
3.
4.
5.
6.

Circle the matching shape.

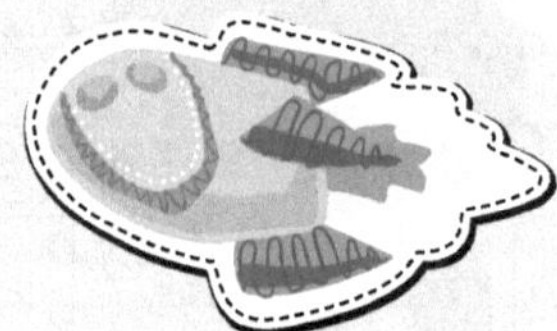

1.

2.

3.

  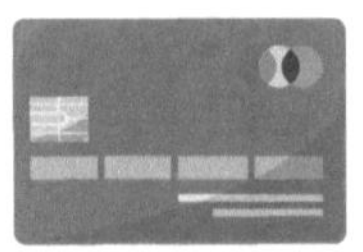  

4.

5.

6.

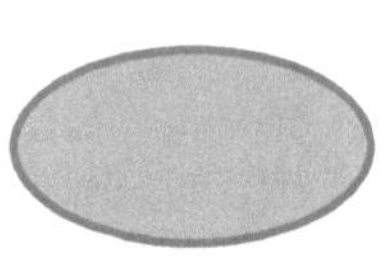    

Week 8 Handwriting Review

Topic 1 Letters Pp and Qq

Today you are going to review and practice printing the letters **Pp** and **Qq**.

Remember Pp is for prince and Qq is for queen.

Directions: On the lines below, practice printing the letters in your best handwriting.

P ....

p ....

Q ....

q ....

FITNESS PLANET

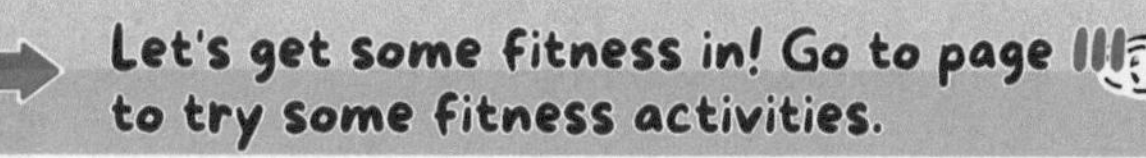

Let's get some fitness in! Go to page 111 to try some fitness activities.

Week 8 Phonics Review

Topic 2 Sound Fluency

If you want to be a fluent reader, or someone who can read well, you need to know letter sounds. It is important to read and say the sound for each letter of the alphabet.

Today, you are going to practice making the sounds for each letter.

Directions: Ask someone at home to help you with this lesson. Look at the table below. Say the sound for each letter in the table. Practice this several times and see if you can go faster each time.

a	n	f	d	t	i
d	f	r	x	c	y
q	m	b	p	e	j
h	a	e	w	o	g
s	v	l	h	s	q
v	k	c	m	b	u
p	r	o	g	t	l
i	u	n	z	j	k

Topic 1 Physical Features of Plants

The **physical features** of a plant are the characteristics that you can observe using your five senses.

Today, you are going to observe and list the physical features of a plant.

Directions: Using a plant found either in your house or outside in nature, observe and list its physical features in the table below.

What does it look like?	What does it smell like?	What does it feel like?	What does it sound like?

Draw a picture of your plant below. Be sure to include any leaves, seeds, stem or flowers that can be observed.

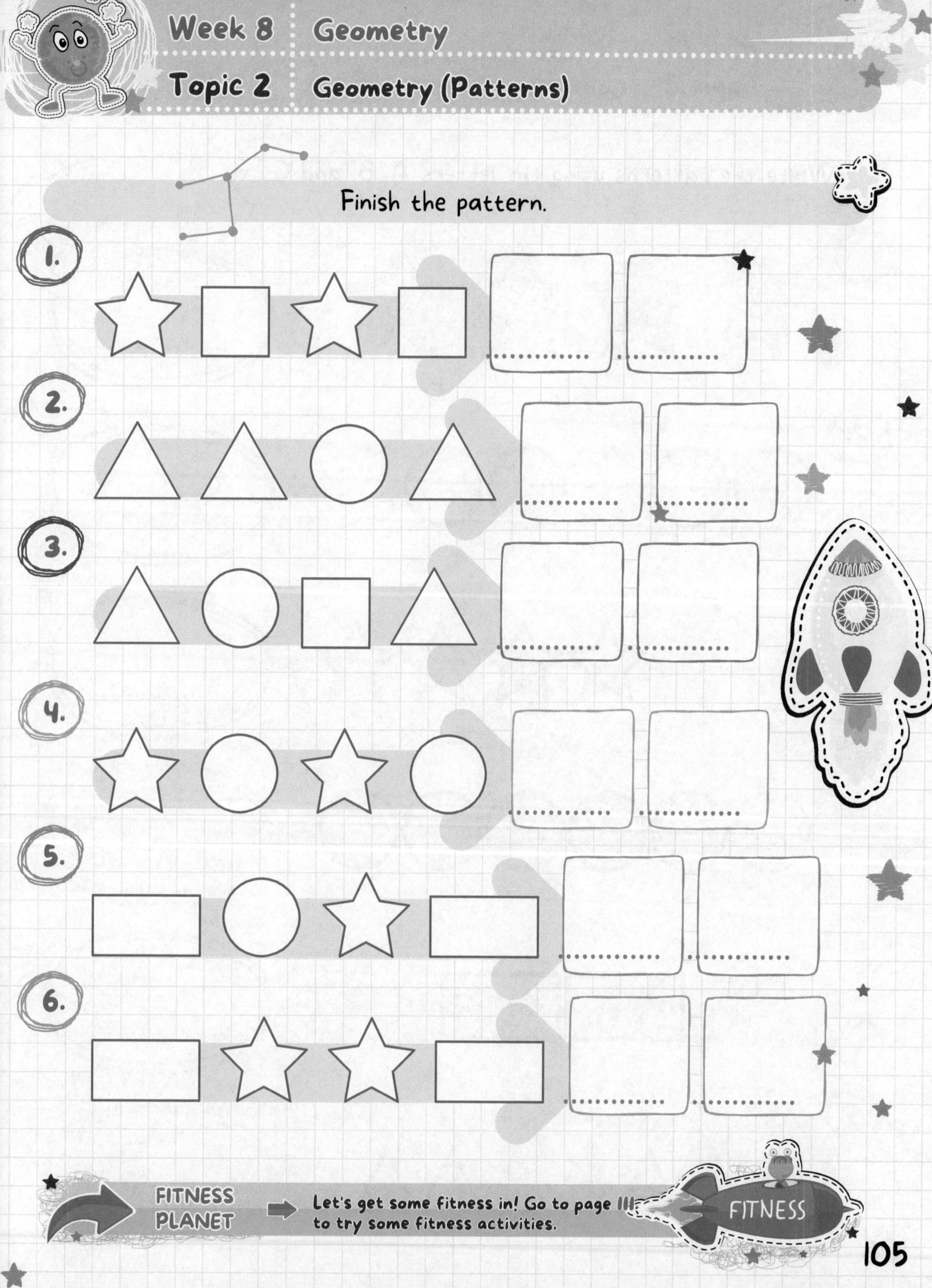

Finish the pattern.

1. ....

2. ....

3. ....

4. ....

5. ....

6. ....

FITNESS PLANET → Let's get some fitness in! Go to page 111 to try some fitness activities.

Write the patterns using the letters A, B, and C.

1. → ....

2. → ....

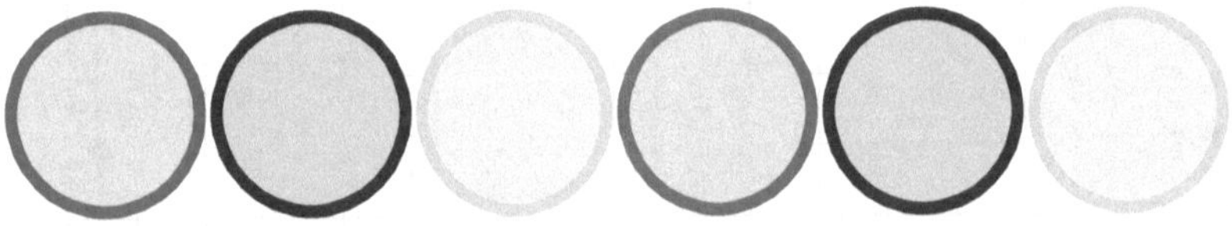

3. → ....

4. → ....

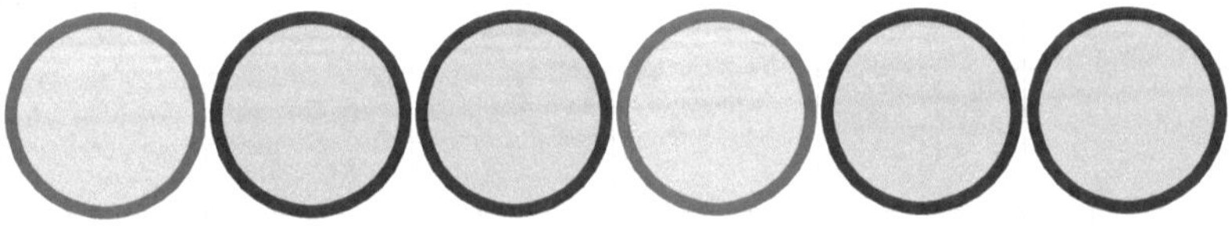

5. → ....

6. → ....

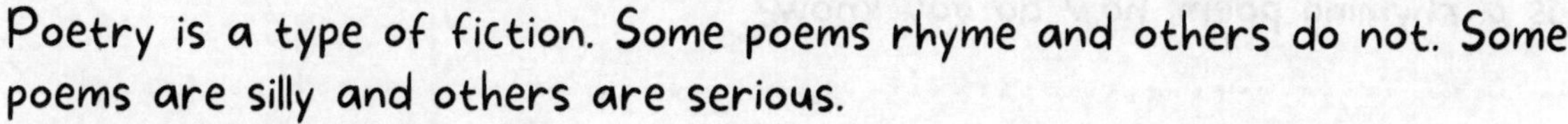

Poetry is a type of fiction. Some poems rhyme and others do not. Some poems are silly and others are serious.

Today, you are going to read a poem and answer questions about it.

The Rat and the Cat

Pat the rat is being chased by a cat,
So he runs across the room and hides behind a broom.

The fat cat looks around for Pat the rat,
He sniffs here and there and everywhere.

Pat thinks he's alone and then lets out a groan,
The cat begins to bat at the handle of the broom.

It's time to zoom across the room,
He sees a hole so that's his goal!

Through the hole Pat begins to roll,
Then SPLAT goes the cat as Pat the rat gets away.
He'll have to catch him another day.

Directions: Now answer the questions about the poem.

1. Is this a rhyming poem?

A. Yes
B. No

1. If it is a rhyming poem, how do you know?

2. List 5 sets of rhyming words from the poem.

3. Why did the author write this poem?

A. To teach the reader about something
B. To entertain the reader

4. Write 3 words that rhyme with "sing."

5. In the poem, "zoom" probably means:

A. To walk slowly
B. To run quickly
C. To roll

6. What do you think happened to the cat at the end of the poem when it says, "SPLAT goes the cat?"

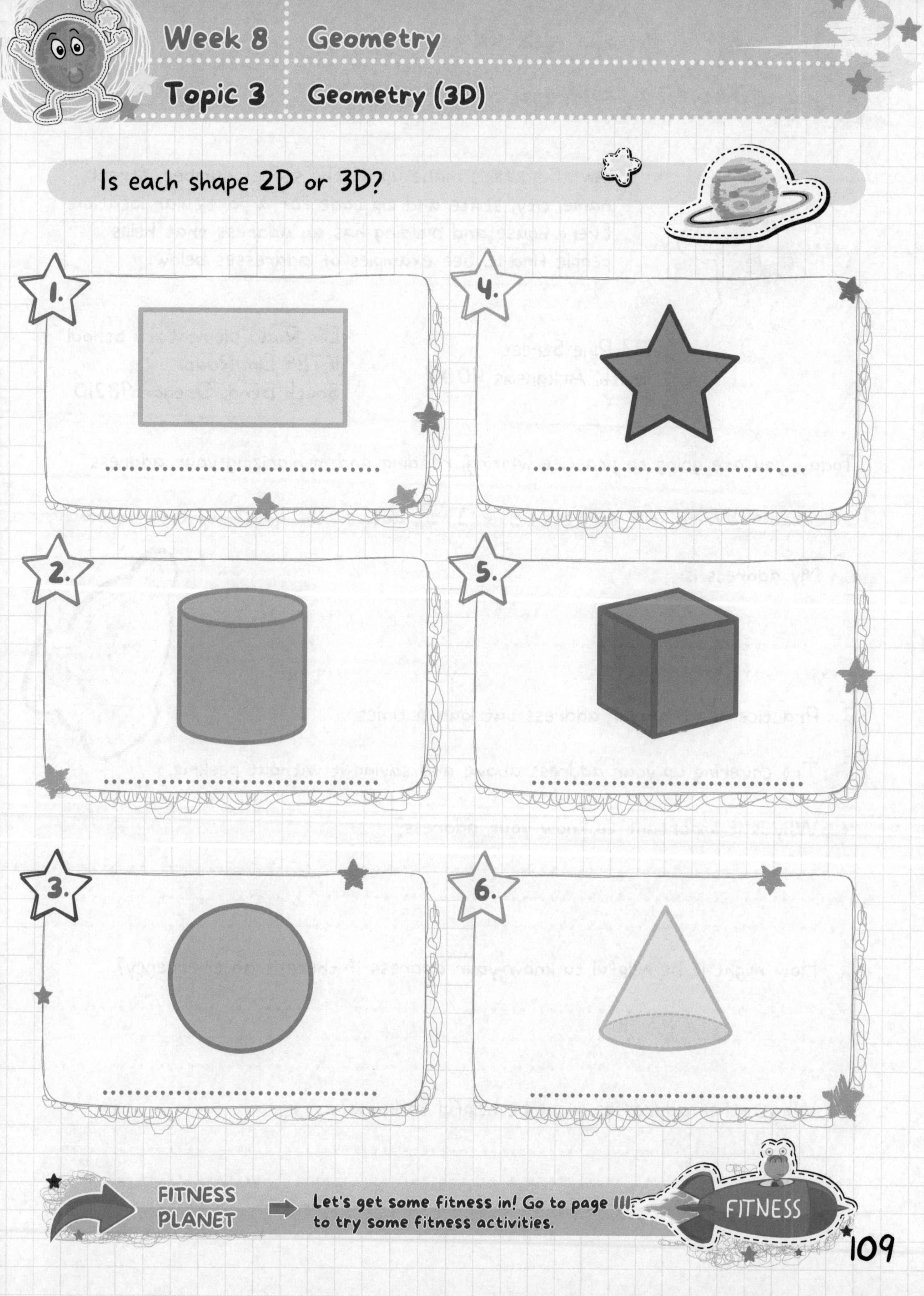

Is each shape 2D or 3D?

1. ....

2. ....

3. ....

4. ....

5. ....

6. ....

FITNESS PLANET

Let's get some fitness in! Go to page 111 to try some fitness activities.

Week 8 Social Studies

Topic 1 Addresses

An **address** is made up of the street number, street name, city, state and zip code for a particular location. Every house and building has an address that helps people find it. See examples of addresses below:

123 Pine Street
Smith, Arkansas 40017

Elm Road Elementary School
16789 Elm Road
South Bend, Oregon 98210

Today, you are going to practice writing, reading and memorizing your address.

Directions: Write your address below. Then, answer the questions that follow.

1. My address is:

....

....

....

2. Practice reading your address out loud 3 times.
3. Try covering up your address above and saying it without peeking.
4. Why is it important to know your address?

....

....

....

5. How might it be helpful to know your address if there is an emergency?

....

....

....

6. What other addresses might be helpful to know?

....

....

....

Kindergarten

FITNESS PLANET

I ♡ FITNESS

I ♡ FITNESS

I ♡ FITNESS

FITNESS PLANET

Repeat these exercises 2 ROUNDS

exercises complex one

2 - **Lunges:** 2 times to each leg.
Note: Use your body weight or books as weight to do leg lunges.

1 - **Abs:** 3 times

3 - **Plank:** 4 sec.

4 - **Run**: 20
Note: Run 10 meters to one side and 10 meters back to the starting position.

Please be aware of your environment and be safe at all times. If you cannot do an exercise, just try your best.

exercises complex two

1 - **High Plank:** 4 sec.

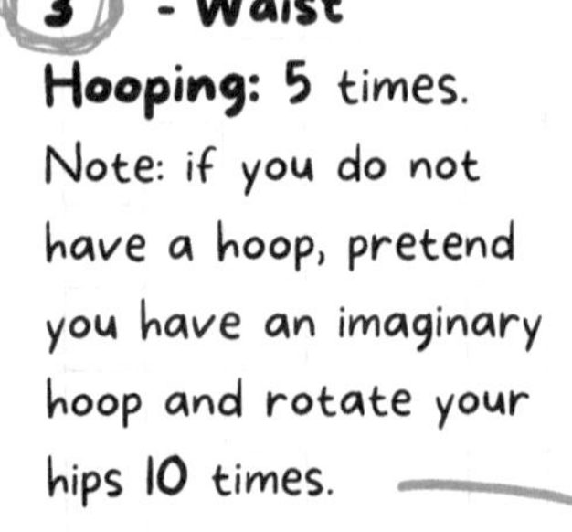

3 - **Waist Hooping:** 5 times.
Note: if you do not have a hoop, pretend you have an imaginary hoop and rotate your hips 10 times.

2 - **Chair:** 10 sec.
Note: sit on an imaginary chair, keep your back straight.

4 - **Abs:** 5 times

FITNESS PLANET

Repeat these exercises 2 ROUNDS

exercises complex three

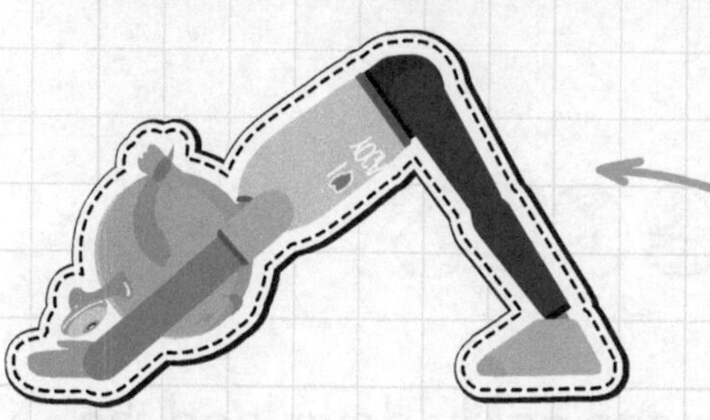

1 - **Down Dog:** 5 sec.

2 - **Bend Down:** 5 sec.

3 - **Chair:** 5 sec.

4 - **Child Pose:** 20 sec.

5 - **Shavasana:** as long as you can. Note: think of happy moments and relax your mind.

Please be aware of your environment and be safe at all times. If you cannot do an exercise, just try your best.

exercises complex four

1 - **Bend forward:** 10 times.
Note: try to touch your feet. Make sure to keep your back straight and if needed you can bend your knees.

2 - **Lunges:** 3 times to each leg.
Note: Use your body weight or books as weight to do leg lunges.

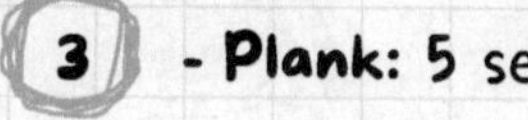

3 - **Plank:** 5 sec.

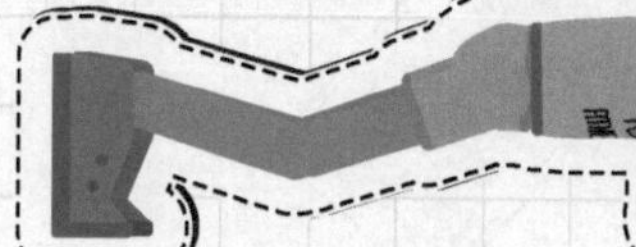

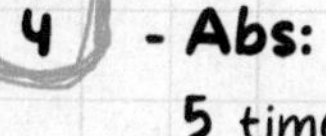

4 - **Abs:** 5 times

Answer Sheets

To see the answer key to the entire workbook, you can easily download the answer key from our website!

*Due to the high request from parents and teachers, we have removed the answer key from the workbook so you do not need to rip out the answer key while students work on the workbook.

Go to **argoprep.com/brainsummer1**
OR scan the QR Code:

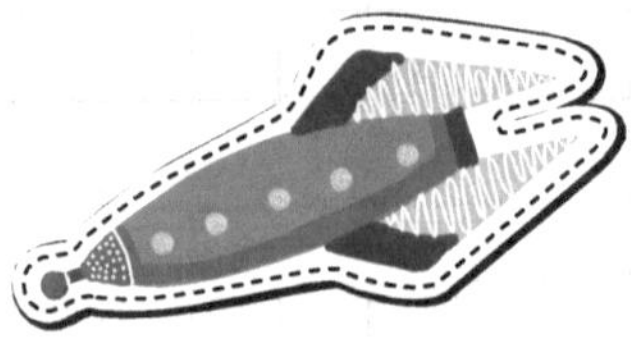

Place your mouse over the workbook you have, and you will see the "Download Answers" button.

Made in United States
Troutdale, OR
05/06/2025